HAND BOOK OF
PIG FARMING

HAND BOOK OF PIG FARMING

Written By :

EIRI Consultants and Engineers

ENGINEERS INDIA RESEARCH INSTITUTE

4449, Nai Sarak, Main Road, Delhi-110 006 (India)

Ph. : 91-11-23918117, 23916431, 23840361

Fax : 91-11-23916431, 23918117

✶ **E-Mail :** eirisidi@bol.net.in

✶ **Website :** www.eiriindia.org

Sole Distributors :

INDIAN INSTITUTE OF CONSULTANTS,

4/27, Roop Nagar, Delhi-110007

Printed and Published by :

ENGINEERS INDIA RESEARCH INSTITUTE

4449, Nai Sarak, Main Road, Delhi-110 006 (India)
Ph. : 91-11-23918117, 23916431, 23840361
Fax : 91-11-23916431, 23918117
✶ **E-Mail** : eirisidi@bol.net.in
✶ **Website** : www.eiriindia.org

Distributed by :

SMALL INDUSTRY RESEARCH INSTITUTE
INDIAN INSTITUTE OF CONSULTANTS

ISBN : 81-86732-85-3
Printed in New Delhi (India)

Rs. 400

Printed and Published by Sudhir Gupta for "Engineers India Research Institute", 4449, Nai Sarak, Main Road, New Delhi-110 006 and Printed at Swastik Offset, Delhi

Preface

*The book '**Hand Book of Pig Farming**' covers Pigs, Swine Rearing in India, Breeding and Breeding of Pigs, Feeds and Feeding of Pigs, Management Practices for Pigs, Health and Hygiene of Hogs, Pork and Pork Products, Plant Economics of Pig Farming, Commercial Pig Farming, Plant Economics of Piggery Meat Processing. Suppliers of Plant and Equipments etc.*

The book has been written for the benefit and to prove an asset and a handy reference guide in the hands of new entrepreneurs and well established farmers.

Director

ENGINEERS INDIA RESEARCH INSTITUTE

Nai Sarak, Main Road,
Delhi-110 006 (India)
E-mail : eirisidi@bol.net.in

CONTENTS AND SUBJECT INDEX

1

Pigs

The pig is one of the most efficient feed converting animal species. Among the meat producing livestock, it is the only litter bearing animal having the shortest generation interval and faster growth rate. These biological advantages give pigs an important role in tiding over the deficiency of animal protein. Domestication of pigs (*Sus scrota*) in Europe took place somewhere around the Baltic sea in the Neolithic times by crossing domestic with wild pigs. In Asia, pigs (*Sus vittatus*) were separately domesticated in China around 3000 BC from native pigs of middle and eastern Asian mainland which included western India to central China and some of the islands like Japan, Sumatra and Borneo.

The pig rearing in India has been a traditional occupation of weaker sections of rural society. The non-descript indigenous pig which is black to blackish grey in colour and grow good quality bristle and convert human and kitchen wastes into meat, has been the prefrence. Except for few imports of superior quality pigs of exotic breeds by some missionary organizations in early years of the last century. no concentrated efforts were made to improve pig production in India. During the second and third Five Year Plans, however, a coordinated programme for piggery development was taken up in some states. The scheme involved establishment of bacon factories, regional pig breeding stations and pig breeding farms/units and piggery development blocks. Some exotic breeds of pigs, viz. Landrace, Large White Yorkshire. Tamworth and Hampshire were introduced at different pig breeding farms. The major objective was to acclimatize the exotic breeds for distribution to pig breeding units for further multiplication and in use for upgrading the native pigs. Currently, there are 237 pig breeding farms in the country, the largest among the livestock species. Of these 65 are in the north eastern states.

The AICRP on Pigs was initiated in 1971 with four centres in the country to study the performance of exotic breeds. The centres at Indian Veterinary Research Institute, Izatnagar and Assam Agricultural University, Khanpara has Landrace; Jawaharlal Nehru Krishi Vishwa Vidyalaya, Jabalpur and Acharya NG Ranga Agricultural University. Triputi, had Large White Yorkshire pigs. In the second phase, native pigs were introduced to study their performance under optimum input and management conditions and to compare their performance with crossbeds having 50 and 75% exotic inheritance. The project is continuing and has been further strengthened with the addition of two more centres, one each at the Tamil Nadu Veterinary and Animal Sciences University, Chennai, and Kerela Agricultural University, Trissur.

Major concentration of pigs is in the rural areas and more than 80% are native pigs. Pig carcass in India averages around 35 kg as against 76 kg in China, 84 kg in USA, and 90 kg in Germany. The emphasis therefore has to be on improvement of indigenous pigs both through selection and crossing them with exotic breeds depending upon the choice of breed type for the region. Indigenous pig breeds should be screened through extensive surveys to identify types with high prolificacy and better growth. The focus should be on improving reproductive efficiency, feed conversion efficiency and achieving higher body weights. Most of the north-eastern states and tribal areas around Ranchi, Chhotta Nagpur in Bihar and Chhatisgarh areas of Madhya Pradesh prefer black pigs. There is, however, no colour preference in south, north west and western parts of the country. Areas around metropolitan cities prefer exotic or crossbred pigs. There is a great demand of pork in north eastern states and southern states of Kerala, Tamil Nadu, Karnataka and Andhra Pradesh.

As a consequnce of various research and development efforts, pig husbandry and pork production has gained impetus during the recent past and the concept of pig farming is changing from that of a zero input enterprise to that of a semi-commercial one. This is due to the realization of its positive qualities like short generation interval, higher growth rate, higher litter size at weaning, yield of around 2 crops per sow per year, ability to convert efficiently ago-industrial and grain byproducts into meat, etc. Although, pig meat production went up from 0.12 million tonnes in 1982 to 0.42 million tonnes in 1995, it constituted only around 10% of the total meat production in the country. Apparently, the species is not being fully exploited taking into consideration its larger growth and prolificacy potential.

Population trends

India with pig population of 12.8 million in 1992 formed only 1.5% of the total world pig population. Majority of these pigs (nearly 84%) are indigenous and the remaining 16% crossbreds. The population of rural and urban pigs is 91 and 9% respectively. Ten Indian states had more than 0.5 million pigs. Uttar Pradesh had the largest pig population (2.9 million) followed by Assam (1.4 million), Bihar (1.1 million) and West Bengal (0.96 million). The north-eastern states had pig population of 3.8 million.

Eastern region of the country comprising Bihar, Assam, Manipur, Meghalaya, Mizoram, Nagaland, Orissa, Sikkim, Tripura and West Bengal had the largest pig population (4.16 million) followed by southern and northern regions with 3.15 and 1.83 million pigs, respectively. Pig population per 1.000 persons comes to 20, 14.2, 12.3 and 3.7, respectively in eastern, southern, northern and western regions of the country.

BREEDS

Indigenous pigs

Indigenous or *desi* pigs are small in size, pot bellied, hairy and black in colour with elongated face and short ears. Their production capacity is very low, average litter size being 5.2 at birth and 3.0 at weaning. Birth weight ranges from 250 to 700 g. The desi pigs have early age at maturity and first farrowing, shorter inter-farrowing interval and higher bristle production.

Exotic breeds

Many exotic breeds of pigs were introduced in the country during the seventh and Eighth Five Year Plan. However, the importance of selecting only a few breeds accordingly to agro-climatic conditions have recently been understood. The following breeds are, by the large, now being used.

Hampshire : The breeds was originally developed in the USA from the hogs imported from the UK. It is black with a white belt around the chest. The body is elongated, well muscled, clean cut, firm and solid with erect ears. Hampshire sows, under well managed conditions, produce up to 2 litters per year with an average litter size of 8 at weaning. This speaks of good milking and mothering ability of the breed. The sows are very docile. A mature boar weighs around 150-250 kg and a sow 90-200 kg. On slaughter, they give around 80% dressed out ratio. The breed, has adapted well under Indian conditions and is being extensively used.

Large White Yorkshire : This is a native breed England and is used extensively for bacon purposes. It is a large in size with a long and slightly dished face. Body is covered with fine white hairs free from curls. Skin is pink and free from wrinkles. It has a long moderately fine coat. Ears are erect and the snout is of medium length. Litter size ranges from 10-14. A mature boar weighs 300-400 kg and sow 230-320 kg.

The Yorkshire is a prolific breed. Sows have good milking and mothering ability. Carcass quality, growth rate and feed conversion efficiency are good. It thrives well under different climatic conditions. Average litter size at birth is around 10.5 and that at weaning around 9. However, its use is limited in the country due to its very heavy size.

Middle White Yorkshire : This breed is a cross between Large and Small White Yorkshire breeds of England. It is white, with a short head and upturned dished face wide between ears. In view of its small size this breed has been used in some areas for grading up of the indigenous stock. The breed is hardy, grows fast, gives a good dressing percentage, but is not so prolific as the Large White Yorkshire. Boars, when fully grown, weigh 250-340 kg, while sows weigh 180-270 kg.

Landrace : The Landrace is a breed of (Northern Europe) and is a bacon breed. It is white with black skin-spot freckles. Other distinguishing characteristics are long body, large drooping ears and long snout. The breed is noted for prolific breeding and for efficiency in utilizing feed. In carcass quality, Landrace is an excellent breed for crossbreeding. Mature boars weigh 270-360 kg and sows 200-320 kg.

Berkshire : Berkshire is one of the oldest improved English breeds of pigs and is extensively used. It is a typical pork breed of medium size. The animals are black with a short upturned nose. The face is usually dished andears are erect. Good Berkshire pigs grow to market size in 6 months of age. Body weights of mature boars and sows range from 272-385 kg and 204-294 kg respectively. Average litter size at birth is 9.9 and at weaning 8.4. The breed has been suitable under Indian conditions.

Tamworth : The breed derives its name from the Tamworth town of England. It is a very hairy breed with golden red pigmented skin and body coat. Head is long and narrow with a long snout and erect ears. It has a strong back and thin shoulders. The carcass produces bacon of best quality due to its lean body free from excessive fat. Mature boars weigh 200-360 kg and the sow 180-320 kg. Average litter isze is 8.9 and 8.0 at birth and weaning, respectively. The breed has the potential to contribute usefully if exploited under commercial cross breeding programme. The breed is, however, less prolific compared to other breeds.

Large Black : This is one of the oldest pig breeds in the Great Britain. Body is complete black. Ears are long, thin and droopy. Ears normally cover both the eyes for which it cannot see side ways. The breed is renowned for its hardiness and docility. Neck is long and clean with wide and deep chest. Body coat is fine and soft with moderate quantity of straight, black and glossy hair. Sows are very good milkers and raise good crop. Milk yield is so good that the health of the sow is affected while nursing a large litter. It is advisable to wean the litter little early to protect sows health. Litter size averages 10.4 at birth and 9.0 at weaning.

Saddle Back : The breed was developed in Hampshire, England. It is essentially a bacon breed. It adapts well under different climatic conditions. It is more or less like Hampshire breed. It is black in colour with white forelegs and has a compact body with a fairly long head. Snout is straight and ears are short and erect having a forward pitch. The breed is little heavier than Hampshiere and is known for its prolificacy. The breed is said to have inhabit vigor not possessed by any other British breed. Feed conversion efficiency is very high. The breed was once used extensively in the country. However, pedigreed stock is limited now.

Duroc : The breed has its origin in the USA. It is red in colour with shades varying from golden to very dark red. It is a large breed with excellent feeding capacity and prolificacy. The sows are good mothers. The weights of mature boar is 410 kg and sow 250 kg.

SELECTION OF BREEDING STOCK

For maximising production efficiency of pig breeding enterprise, it is essential fo have large number of piglets born and weaned per litter, preferably 2 farrowing per year, higher growth rate and feed conversion efficiency and better carcass quality and yield as per the market demand.

Breeding stock is very critical to ensure higher profits and therefore, its proper selection is important. First, a prudent decision on the type of breed to be raised has to be taken based on its source and market demand. Secondly, breeding stock from within the breed are to be selected preferably on individual merit. While selecting the stock, emphasis may be laid on phenotypic and genetic make up and other important traits. Normally traits like litter size at birth and weaning, weaning weight, feed conversion ability (individual and litter mates), body conformation, genetic defects, etc. are considered for the animal to be selected. In addition, information on temperament, mothering ability, ancestry and fertility of the sows that produced them is also taken into consideration. It is always better to select animals from a herd that has adapted itself in the area and is well known for its productivity.

Number of breeding stock to be procured shall depend on the infrastructure/capital strength of the producer and the feed source availablity. The stock may have a male : female ratio of 1:4 in small holding and 1:8 in large enterprise. Once the herd is established, say with a 50 sow unit, provision for additional infrastructure be made from second year to increase the strength with the multiplied stock. It is easier to select the replacement stock from within one's own herd as the records are available. While selecting replacement stock, it is advisable to select breeding boars and gilts from different lines to avoid inbreeding.

Selection of boars

Selection of a breeding boar is of paramount importance as it is the boar that is to determine to a large extent the quality of the piglets to be produced. In advance countries, some standard norms are fixed for the selection of both boars and sows. For example, a gilt or boar should weigh around 90 kg at 6 months with a body length of 100 cm and a litter weight at weaning of 120-150 kg. In India, there is no breed that will give such performance even under best managerial system. Therefore, our standards have to be little different persently.

Selection of the breeding males at weaning stage itself is important. Males should be selected initially from a weaner lot having higher litter size with uniform growth of its litter mates. Males having a weaning weight of 10 kg and above may be initially selected. Body weight of such selected males should again be taken at 120 days to ascertain its post weaning gain and the ones showing a gain of around 250 g and above per day may be further screened. Body conformity as per breed type and development of teats should be observed. Such selected boars should then be put under uniform feeding and the ones showing maximum weight at 180 days may be selected for breeding purpose. Their libido and overall activeness/alertness should form final basis of selection.

A farmer while buying a breed boar, should first collect the information on its ancestry, performance of its litter mates, genetic defects, etc. The boar to be purchased should be observed from the close distance allowing it to walk up and down to ascertain its conformity, gait, any defect in legs and feet etc. Boars with inverted teats or only with 4 pairs of teats should not be selected. Similarly testicle size and numbers need to be observed.

Selection of gilts

Gilts or the sows bred large litters of strong, healthy pigs and, therefore, their effective selection is extremely important. Gilts should normally be selected from a herd where uniform feeding is practised. Gilts to be selected should possess all the characteristics of the breed. She should be long and deep to carry well grown litters. She should not be bellied or with arch back conditions. Her shoulders should be fine, well set; ribs well sprang and the underline straight with at least 12 functional teats. The jowl should not be hanging down heavily with excess fat. She must walk well and free from any deformity. Temperament of the gilt should be observed. A bad tempered gilt may not look after her offspring well. If these qualities are satisfactory, other information like its ancestry, performance of the dam and half-sisters may be collected. Body weight of the gilts, under Indian conditions should be between 30 and 35 kg at 6 months in *desi*, 45-50 kg in crossbreds and 60-70 kg in exotic breeds.

Systems of breeding

Main purpose of breeding is to produce individuals with superior merit in terms of production, reproduction and overall economics through recombining the genes into more desirable groups. Different methods of breeding are practised depending upon the aim of the breeder.

Breeding systems are mainly of inbreeding (mating of related pigs) and outbreeding (mating of unrelated pigs) types.

Inbreeding is further divided into (a) close breeding, and (b) line breeding depending on the closeness of the relationship. Close breeding is the mating of brother-sister, dam - son, sire - daughter etc. Line breeding is the mating of cousins and other relations. In short, mating of individuals with 4-6 generations may be termed as inbreeding. This system of breeding, though desirable to evolve genetically superior animals, be left to be handled by the breeder as it involves equal amount of risk leading to inbreeding depression.

Outbreeding has 4 different types, viz (a) outcrossing, (b) crossbreeding, (c) species hybridization and (d) grading up. Out of these, crossbreeding has 3 different systems, viz, criss-crossing, triple crossing and back crossing.

Outcrossing : It is the system of mating where unrelated animals within the same breed from different sire/dam lines are selected and bred to produce outcross progenies.

Crossbreeding : **Criss crossing**. It is the mating of two different breeds alternatively, e.g, Hampshire sows crossed with Landrace. Their progenies are again bred wth Hampshire and the resultant progenies with Landrace and so on. In this system, one can produce pigs with different levels of inheritance from the two breeds and find out the best combination.

Triple crossing. This is the system of using 3 breeds in a rotational manner. This is also called rotational breeding. e.g. crosses of Hampshire x Landrace are bred to Large Black, a third breed. Their crosses are bred back to Hampshire and theirs to Landrace in a rotational manner. The crossbreds so produced will have genes in different proportions from all the 3 breeds.

Back crossing. It is the mating of crossbred animals back to one of the purebred parents which were used to produce them.

Grading up : It is the system of breeding purebred sires to non-descript (local) pigs generation after generation. Through this system, level of exotic breed inheritance could be increased in the upgraded stock up to 99% in seventh generation. Eventually, after seventh generation, an upgraded stock similar to the breed used could be produced from the indigenous line.

Among the above systems, outcrossing, criss-crossing and grading up are mostly being used in the country for breeding pigs. Crossbred pigs are superior in terms of productive and reproductive traits than the pure breds that were used to produce them. Similarly, performance of graded pigs was better than indigenous. It will be best to select the procure stablished graded pigs with suitable level of exotic inheritance and to follow out-crossing or *interse* mating (within the group). It may be remembered that selection and breeding are the 2 main tools to produce superior individuals.

Hybrid vigour : When 2 breeds are crossed their offspring sometime show performance superior to either of the parent breed. This is known as hybrid vigour or heterosis and is mostly associated with traits having low heritability (number of pigs farrowed and weaned, litter weight at birth/ weaning, mortality from birth to weaning). Pig breeders have extensively made use of hybrid vigour and produced commercial crosses using specialized selected lines. Criss-crossing and rotational crossbreeding schemes have been used to maintain hybrid vigour in females.

FEEDING AND NUTRITION

Feeding alone accounts for 70-75% of the total production cost. Feeding pigs has, therefore, to be balanced and economical. Two types of feeding are practised: a) concentrate feeding, b) mixing recommended concentrate feed with other locally available agro-industrial by-products including incorporation of tuber crops like sweet potato, topioca and radish beside some other fodder crops. It goes without saying that profit margin will be more with concentrate feed as 1 kg of live weight could be produced with less amount of such feeds. Locally available agro-industrial by-products when mixed in ideal proportions are good for maintenance and finisher pigs. Pig producers are expected to know the following while feeding their pigs: (i) nutrient requirement, (ii) characteristics of a good ration, and (iii) how much to feed?

Nutrient requirements

Nutrient means a single class of food or a group of feeds that aid in support of life and help the animal to produce as per expectation. The nutrients consist of 10 essential amino-acids, 17-vitamins, 13 or more minerals, essential fatty acids, carbohydrates and some unidentified factors.

Carbohydrate : Main function of carbohydrate in animal body is to provide heat and energy and also to provide material for fattening. All animals including pig need more of carbohydrates than all other nutrients combined. Sources of carbohydrates are cereals, viz, maize, oats, wheat, barley, rye, and rice. By-products of grains, wheat bran, rice bran, rice polish, rice meal, etc. are utilized conveniently in pig feeding. Grains and energy feeds do contain some protein, vitamins and minerals but these are two low to provide adequacy.

Pigs unlike cattle, sheep and goat are simple stomach animal. Therefore, they cannot digest more fibrous foods. Cattle and sheep, for example, have a great deal of microbial activities in the rumen (stomach portion) which break down a large part of crude fibres. Pigs cannot do this. Hence, fibrous feeds like hay, etc. should not be fed in high proportions to pigs. In any case fibre content should not exceed 6-8%. Carbohydrate constitutes about 70% of the total feed. Therefore best and economical carbohydrate feeds as available locally should be arranged for feeding pigs.

Protein : In animal feeding protein either builds new tissues as in growth and reproduction, or repairs worn out tissues. Its lack in diet leads to reduction in growth or loss of weight. Protein requirement varies as per the category of pigs. Growing pigs need around 22% protein during suckling period and 18-20% in weaner stage. This requirement is reduced as they grow. Protein requirement of a breeding boar and pregnant female is around 15%. Main sources of protein are oilcakes (groundnut-cake, till oil-cake, soya-flakes) and meals of animal origin (bone-meal, meat-meal, fish-meal etc.)

Minerals and vitamins : Minerals like calcium, phosphorus, iron, copper, cobalt etc. are required in trace amount. Deficiency of minerals will produce several deficiency symptoms. Commercially available minerals mixtures may be added in feed @ 1 kg/100 kg of feed, to supplement mineral requirement. Iron and phosphorus may, however, be added separately as sow's milk is deficient in iron.

Pig feed should contain vitamin A and D (fat soluble) and niacin, riboflavin, vitamin B_{12} and pantothenic acid (water soluble) from the B-complex group. Generally vitamins in adequate amounts are available in a good ration. However, in confinement system of rearing, suplementation of vitamins, except vitamin D which is available from sunlight, may be essential. These may be ensured through the incorporations of vitamin mixtures.

Salt is also an essential item in pig ration. Different researchers have, however, varied opinion on the requirement of salt. The range given is from 0.5-5.0%. However, 1.5-3.5% salt in pig ration is by and large recommended. Excessive salt is unpalatable and may be toxic to the pig. It is advisable to use iodised salt to meet the requirement of both salt and iodine. Iodine deficiency often results in the birth of hairless pigs or goiter.

Water : Water is vital for life process. The pigs should be given *ad lib.* water to drink particularly during hot weather. Requirement of water is also more for sows suckling piglets. If wet mash feeding is practised, extra water should be provided after each feeding.

Feed additives : Feed additives are used as anti-microbial compounds and growth promoters. Among the anti-microbial compounds antibiotics are extensively used in pig ration. Antibiotics are not nutrients but they benefit by acting upon the system associated directly or indirectly with microbial flora of the host animal. Antibiotics are drugs and some of the commonly used antibiotics are bacitracin, chlorotetracycline, neomycine, oxytetracycline, penicillin, streptomycin, etc. This is, however, a rapidly changing area and suitable new drugs may keep coming. Feed additives may be added in pig ration to supply the antibiotic. Other feed additives like chelated aminoacids also promote growth rate in pigs.

Characteristics of a good ration

A good ration provides the needed nutrients in such proportion that the animal will be able to convert the same effectively into park. The composition of a ration cannot be static all the times. Market availability of feedstuff and their prices will determine the choice of the ingredients.

A good ration should contain the required nutrients in right proportion and as per the need of different categories of pigs. For example, protein content should be 20-22% in creep ration, 16-18% from weaning to 120 days, 13-15% in growers, 15-16% in breeding boars and pregnant sows and 12-14% in finishers/ dry sows ration. Proportion of different ingredients in various diets to meet requirement of energy and protein is given in Table 1.

Table 1. Different ingredients of pig ration.

Ingredients	Per cent as per category				
	Creep	Weaner	Grower	Market	Breeding boar/ pregnant sow
Maize (crushed)	60	55	50	45	50
Wheat bran	14	17	20	25	18
GN-cake	14	16	18	20	20
Molasses	5	5	5	3	5
Fish-meal	5	5	5	5	5
Salt	1	1	1.5	1.5	0.5
Mineral mix	1	1	0.5	0.5	1.5

Feed additives as per recommendations may be added.

It is, however, very important to examine the quality of feed ingredients. Mouldly feeds should be avoided as these might lead to afflatoxicosis, a feedborne disease with a heavy mortality particularly in young piglets. In this condition, the given feed should be withdrawn immediately and replaced with other feed. Similarly, water should be from a clean source to avoid water-borne diseases. Water may be periodically treated with potassium permanganate.

Pigs belonging to different categories have different feed requirements. A broad guideline is given below:

Weaner pigs	56 - 120 days	:	250 - 750 g/day
Grower pigs	121 - 180 days	:	750 - 1.50 kg/day
Adults	180 - 300 days	:	1.50 - 2.50 kg/day
Sow		:	2.50 - 3.50 kg/day

In concentrate feeding, it is better to purchase the ingredients and compute one's own ration. Addition of feed additives check infection and ensure higher growth. Similarly addition of mineral mixture and sometimes the aminoacid besides vitamin is recommended. Under Indian conditions, wet mash feeding is preferably. If suitable alternatives are found, requirement of concentrate feed could be reduced. It is very important to ascertain the feed conversion feed could be reduced. It is very important to ascertain the feed conversion efficency in the hard. This will indicate if any change in the ration is necessary. Expenditure incurred on feed must be recorded. During summer, the pigs may be given some grasses/fodder, waste vegetables etc.

MANAGEMENT

Management plays a key role towards the profitability of a piggery enterprise. Management practices to be followed for various categories are discussed here.

Boar

Breeding boars need extra care. They should preferably be fed individually with measured quantity of feed. Normally a breeding boar between 8 and 12 months of age may be given 2-3 kg of feed per day with *ad lib.* water. It is the feeding time when a boar should be observed for any ailment, lethargy, etc. Slippery floors for boar may be avoided as this might lead to injury of the legs. The boars should not be allowed to sleep immediately after feeding as overfed boars will tend to sleep and, in turn, accumulate fat. Special ration (flushing) therefore should be followed at this stage with sufficient exercise. Each boar mey be given a floor space of 60-100 sq.ft.

Not more than 4 services per week per boar should be allowed. Breeding time should be confined to either early or late hours of the day. A service crate is best for breeding. Individual breeding record may be maintained to asses the performance of the boar. At no time, however, two boars are released at breeding time as this will lead to severe fighting. General health examination of the boar including the screening of ecto- and endo-parasites be carried out periodically.

Sow

After the pregnancy is diagnosed (expert managers can diagnose pregnancy visually after one month of breeding) all care should be taken for the sow to ensure a better harvest. No pregnant sow should be shifted/ taken to a new house where others may treat her as an intruder because in pigs general tendency is to fight with any new comer. Pregnant sows should be fed individually. Each sow, however, towards the end of third month of pregnancy, should be shifted to a farrowing room having a floor space of 80-100 sq. ft. covered area and an equivalent open area. Provision of a bedding be ensured after she crossed third month of pregnancy. Feed quantity needs to be reduced from around 107 days of pregnancy.

Sows normally do not need any assistance during farrowing. As the labour pain starts, she should not be disturbed in any way. Only the familiar pigman may attend to her. As the delivers, the new born piglets should be cleaned and helped to the teats. Naval cord of the piglets should be cut and the expulsion of the placenta ensured. If placenta is not expelled up to 4 hr from the delivery of last piglet, the uterus of the sow should be examined for any more piglet. If not, the placenta should be removed manually. Each piglet should be examined carefully while on teats and the weaker ones helped to the teats to ensure that each one of them got the colostrum. From second day, the feed of the sow may be increased gradually. Normally, an addition of around 100-150 g of feed per suckling piglet over and above the normal quantity for the sow helps in maintaining her health. Temperament of the sow, her mothering ability, cannabolism, etc. should be observed to take to remedial measures. If the litter is weaned within one month, the sow could be rebred within a week.

Pre-weaner

Young piglets are devoid of body hair and have less subcutaneous fat for which heat loss from the surface is very high. They, therefore, need a good housing. During winter, provision of a heat lamp is necessary as the temperature difference between the sow's womb and the environment is around 20°F. Otherwise also, provision of a light inside the creep area will help in attracting the piglets away from the sow. This helps to check the mortality due to overlaying by the sow. Iron (ferrous sulphate) may be applied to the teats periodically so that the piglets do not suffer from anaemia. 'Inferon' injection at this stage is better. Exposing the piglets to soil also helps to get some iron from it. Bedding with saw dust or paddy straw wrapped in a gunny bag is preferable.

When the piglets are around 16-20 day old, they may be given creep ration in addition to sow's milk. This ration is given inside the creep/bar area so that the sow does not have access to it. The piglets may sometimes suffer from scour (white diarrhoea). In that case the sow may be treated with neftin. Other hygienic measures should be ensured to raise as many piglets as possible to weaning stage.

Weaner

Weaning is a crucial stage for the pigs because at this stage they are separated from their mothers and have to support themselves. Normal weaning period is 56 days. However, efficient managers have been successful to wean the piglets as early as 21 days. With 21 days weaning, it has also been possible to get around 2.2 farrowing per sow a year. This, however, needs higher managerial skills and knowledge on piglet feeding.

Each piglet should be given an identification mark at weaning so that their ancestry is well documented. There are different methods of identification including ear-notching. One of the simplest methods is to give a number by cutting its body hair. In this method, however, the number will have to be cut repeatedly after every two months as the hairs keep growing. This, together with the recording of other distinguished body features is the simplest of the methods. Piglets should be weaned in groups so that they do not develop the stress of isolation. Within a period of one week, they may be deworned. Periodic examination of faecal samples is recommended. Wet mash feeding at this stage is preferred. As they grow, males and females are separated.

Proper records should be maintained and vaccination against known infectious diseases like swine fever be carried out. Daily and routine management charts may be prepared indicating the date/month for vaccination, deworming, weaning, identification, breeding, flushing, marketing etc. and these activities carried out as per schedule.

General

It is important to clean all the rooms every day before giving morning feed. Each pig should be observed during feeding time to identify the sick animals or animals in heat. These are then to be recorded and suitable treatment/breeding plan prepared. Maintenance of herd strength, breeding, feeding, disease and expenditure record registers is considered a must to work out the balance sheet/profits.

Culling

Culling/elimination of unproductive pigs from the herd should be done regularly. It is futile to keep feeding the unproductive or less productive animals. Sows after 6 farrowings may be culled as the litter size normally declines after this. She should be replaced with another from the herd itself. Gilts that fail to conceive after 3 services and the sows with very high inter-farrowing interval period should be eliminated from the herd. Similarly excess and surplus boars need to be culled when they are around 6 months of age. Lame or other perpetually debilitated animals may also be culled. Pigs having atropic rhinitis, cryptorchidism, etc. should also be disposed off. Such practices will ensure better profitability from the pig enterprise.

Labour management

Next to feed, the labour accounts for second highest expenditure. In case of a pig farm, labour to pig ration may be kept at 1:40 with 2-3 extra labourers who may be engaged in farm duties whenever needed. Duties of the farm supervisor (s) and stockman etc. should be clearly spelt out and a logbook maintained. Incentive/bonus for any additional profit may be given. Personnels deployed for the purpose marketing should be managed separately.

Pig housing

Housing requirement for pigs vary with its category. A breeder may, therefore, plan to have houses for weaner, grower, boar, pregnant, lactating and dry sows. The houses should be so arranged that shifting the animals from one house to the other becomes easy. Like next to weaner house, grower, boar, pregnant and farrowing houses should be constructed so that the animals could be shifted in a rotational manner.

Type and design of the pig sty will vary according to the location and availability of building materials. In the high altitude areas, heights of the house may be kept low and the roof made little sloppy so that heat losses from the surface are less and the rain water is drained off easily. Height of the sheds in plains may be little more with enough space for wind movement.

Normally a double row system covered area with a passage in between is ideal. The covered area will have the provision for feeding and water troughs so that the pigman giving feed may serve pigs on either side of the passage with comparative ease and efficiency. Each room in the covered area should be linked at the back side with an open area. Size of each room could be 10' x 10 x with the provision of equal amount of space in the open area i.e. a total room size of 10' x 20'. This space could accomodate either 12-15 weaner or 8 grower/4 dry sows/2 pregnant sows/2 boars. Eight such rooms on either side of the row may be had under one roof. Separate provision for farrowing pens shuld be made. The farrowing pens could be 80-100 sq.ft. depending on the breed type to be kept. The farrowing pens should have the provision, in addition to feeding/ watering troughs, for a guard rail. The guard rail helps in free movement of piglets and is also used for creep feeding. In the high altitude areas, farrowing pens should also have the provision for heat lamps (bulbs) in the guard rail area.

Floor of the houses should normally be little rough to avoid slipperiness. The floor may also be made little sloppy for easy drainage of urine/dung etc. Each room of the sty may be linked to a common drain and all drains linked to a dung pit.

Tubular structures for the roof, though costly compared to wooden beams are durable. Partition walls in between the rooms may be made with bricks up to around 3-4 feet. G.I. pipes, in place of bricks, may also be used in the houses other than weaners. Fattener farmers particularly in the hill areas have developed some cheap and effective housing designs. Locally available housing materials like bamboo, jungle wood post and thatched grasses are normally used. Floor is made of bamboo post on the sloppy terrain of the hills. This ensures easy drainage of excreta into the natural slope. Feeding trough is made of used cut tyres. Roof is constructed with thatched grass and the walls with jungle wood post. This type of housing is suitable only in hill areas where the farmer rears 3-4 pigs for fattening. Floor space allowance (Table 2) is an important aspect to be kept in mind while constructing houses for different categories of pigs.

Table 2. Floor allowance for different categories of pigs

Category	Covered space/pig (sq. ft.)	Open space/pig (sq. ft.)	RH (%)
Weaner	10 - 15	15 - 20	60 - 80
Grower	12 - 20	20 - 30	60 - 80
Boar	35 - 50	50 - 70	60 - 80
Lactating sow	70 - 100	70 - 100	60 - 80
Dry sow	20 - 30	30 - 50	60 - 80

Different designs for pig sites for different areas are available. One should ensure judicious use of pig sty for economic reason. Good pig housing has been found to contribute up to 20% towards net production.

Marketing and transport of pigs

Marketing of pigs is either done as breeding stock or as fattener. Breeding pigs are either sold at weaning or as adult breeding boar or sow. Since quality pigs are not available, the breeder preferably sells his pigs at weaning. The efforts should be to obtain an average weight of 12-15 kg at weaning and 60 kg at 5-6 months.

Pigs for slaughter are transported to bacon factories or urban markets. Indigenous pigs invariably are made to drive on foot. An exotic pigs cannot travel long distances, without losing weight, they are transported in trucks/ rail. In both these cases, pigs are fed en-route and watered. Hot summer and mid-day sun should be avoided and journeys made during the night to provide confort during the movement. The animals should be transported in trucks designed for livestock transport. The weight loss during transport is 2-5% of body weight. Enough care should be taken while loading and unloading the pigs to avoid injury to the animal. It is advisable to build permanent loading ramp built in the farm to avoid any damage during loading/ unloading.

PIG DISEASES AND HEALTH

Swine herd health is an important aspect of overall swine production programme. A sound herd health preventive programme includes good sanitation, vigorous parasitic control and vaccination against important disease. It is essential to develop a data base on pig diseases which will help in evolving suitable control measures. Important prevailing diseases of pigs in our country are tuberculosis, mycoplasmal pneumonia, edema, swine fever, FMD, swine-pox, parasitic diseases, etc.

Tuberculosis

It is a chronic bacterial disease and is highly communicable to human beings. The clinical symptoms are very vague. If the disease is progressive, general symptoms like unthriftiness, nodular swelling of the neck region, digestive disturbance, pulmonary symptoms, enlargement of bones and joints are observed. On postmortem examination, nodules/tubercles may be ssen in lymphnodes, liver, lungs and spleen. No satisfactory preventive and curative measures are available. Regular tuberculin testing and meat inspection of pig carcasses will help in detecting the disease in the population and its further elimination.

Mycoplasmal pneumonia

This is the world's most important swine disease. Economic losses in terms of morbidity and mortality are very high. Pigs between 3 and 10 weeks of age are worst effected. The cough which is most marked when pigs come out to feed in the morning is very characteristics. In general, pigs retain their appetite but do not grow well. Intercurrent infections due to parasites may lead to 25-30% losses.

Colibacillosis and edema

The disease is caused by specific pathogenic types of *Escherichia* coli. This disease manifests in two distinct clinical forms. Diarrhoea with toxemia/septicaemia is commonly encountered in pre-weaning stages. Edema or enterotoxemia is found in post-weaning stages. Morbidity and mortality rates vary between herds and across regions.

Brucellosis

This is mainly a reproductive disease which causes abortion, still birth, infertility, orchitis and lameness. There is no treatment. Positive cases after the antigen test must be removed and not allowed to breed.

Atrophic rhinitis

It is characterized by atrophy of nasal turbinates, nasofacial deformity/ distortion and poor weight gains. This disease was recorded in the north-east hill region. In 6-to-8-week- old pigs, the disease is commonly observed with mild to severe rhinitis accompanied by nasal discharge and nasofacial distortions. Although, morbidity rate is high mortality is low.

Swine fever

It is a highly infectious viral disease of swine characterized by high fever, and generalised haemorrhages. Morbidity and mortality rates are variable according to the status of immunity in a population. Wherever high mortality rates are reported in any outbreak, swine fever should be suspected immediately and suitable measures taken for confirmation of disease. Dullness, depression, anorexia, high body temperative, staggering gait and conjunctivitis are noted.

Foot-and-mouth disease

It is an acute and highly communicable viral disease of cloven footed animals. The disease is characterized by formation of vesicles and erosions in the mouth, on snout, feet, dewclaws, teats and udder. High morbidity rates are commonly observed among the affected populations with variable mortality especially in young animals. There is no effective medicine except use of 2% alum or potassium permanganate for mouth wash and 1% copper sulphate solution for foot lesions. Vaccine protects against one or two strains and that too is limited for 3-4 months.

Swine pox

The disease is caused by swine pox virus and vaccinia virus and is manifested by typical pox lesions on soft tissues of ears, face, neck and teats. Pigs of 2-4 months age are geneally affected. Ulcers in digestive tract and pneumonia are also noted in some cases.

Parasitic diseases

Ascariasis : *Ascaris suis* is a large round worm found in small intestine and lives on food taken by pig. The main symptoms are unthrifty appearance, pot bellied, roughness of coat and stunted growth. Lungs and liver also get affected resulting in difficult breathing. Eggs of the worm are passed in faeces. Sodium fluride, piperazine compound and thiabendazole are used for treating affected pigs.

Metastrongylus infestations : Lung worm with metasrtrongylus infestation affect pigs in the bronchi and broncheoles. The noticeable symptoms are respiratory discomfort and coughing with discharge from nose. The earthworms hosts the infective larvae.

Trichinella spiralis infestation : The adult worm is found in small intestine and its larvae are found in muscles which may be transmitted to human beings. The meat of the infected animals slow white or little brownish pin heads in muscles commonly known as measly pork. Feeding of garbage and meat scraps can lead to this infection in pigs.

Mange : It is a chronic condition caused by *Sarcoptes scabiei* var *suis.* It causes itching and constant irritation leading to restlessness, abrasions of skin and loss of hair. Insecticide spray, dip and ointments are used to cure this disease.

Others

Avitaminosis : Deficiency of vitamins A, B_2 and D_3 are common. Vitamin A deficiency leads to multiple birth defects and night blindless while deficiency of vitamin D leads to rickets. Dermatitis is often caused due to riboflavin deficiency.

Anaemia : This is due to deficiency of iron. As sows milk is deficient in iron, piglets get the anemia. Iron injections are used to cure the disease. Inadequate supply of iron, copper, calcium, phosphorus, etc. pronounce the symptoms.

2

Swine Rearing in India

2.1 INTRODUCTION

Though the practice of swine husbandry exists in India for centuries, its contribution to the national economy is very meagre. Swine rearing in India is carreied out under variety of adverse social, climatic and environmental conditions. Hog Industry has remained undeveloped mainly due to religious taboos and prejudice. Swine rearing in our country is mainly in the hands of socially economically weaker illiterate people from scheduled caste and scheduled tribes. Their unscientific breeding practices and unhygienic management of pigs have also been a contributing factor which kept this industry in primitive stage. Pigs by these people are often reared in small groups and allowed to fend for themselves in open or free range system. There common village hog is a non-descript type scrub animal and has no definite breed characteristics. This is maintained by poorer sections of the society who find difficult to provide land for forage and with no food to eat, so usually they thrive as scavangers. Such a condition of rearing has deprived this animal a fair chance to grow resulting in production of poor quality and quantity of meat (Table 2.1 and 2.2). Thus, local pig rearing has not been a profitable one. To exploit this non-descript hog for large amount of pork, during last two decades a great emphasis has been laid on the improvement of the productivity of pigs by implementing crossbreeding programme with exotic breeds of swine, to develop animals of large sized litter, efficient feed conversion, higher dressing percentage and better qualit pork.

Table 2.1. *Indigenous non-descript (Desi) hog Vs. improved hog of exotic breeds.*

	Parameters	*Desi Pig*	*Improved Pig*
1.	Breed Characteristics	Not definite	Well defined
2.	Body Size	Smaller	Larger

	Parameters	*Desi Pig*	*Improved Pig*
3.	Body Weight	Less	More
4.	Food conversion efficiency	Poor	Better
5.	Growth rate	Poor/Slow	Faster
6.	Maturity	Late	Early
7.	Fattening quality	Poor	Better
8.	Size of litter	Small	Large
9.	Pork yield	Less	More
10.	Dressing percentage	Less	More
11.	Pork quality	Inferior	Superior
12.	Resitant to diseases	More	Less
13.	Sows as mother	Better	Fair
14.	Prolificacy	Low	More
15.	Generation interval	Longer	Shorter
16.	Profitability	Low	Higher
17.	Pork suitability to curing	Fair	Better
18.	Example of Pigs	Local village hogs	Yorkshire,, Landrace, Berkshire

2.2 REASON FOR POOR GROWTH OF COMMERCIAL HOG INDUSTRY.

1. Pigs are not generally used for meat purpose by majority of meat eaters in India.
2. Pigs compete directly with human beings for cereal grains.

Table 2.2 *Comparative performance of indigenous and exotic breeds Pigs in India.*

Traits	*Desi pig*	*Middle white Yorkshire*	*Crossbred*
Percentage of piglets born dead	7.68	7.84	3.87
Mean litter size	7.79	8.71	9.53
Mean birth weight of piglet (kg.)	0.90	1.34	1.23
Mean weaning weeks (kg.)	4.02	7.01	6.65
Mean weight at 48 weeks (kg.)	40.0	73.5	51.0
Growth rate from birth to weaning (kg.)	0.073	0.108	0.102
Feed efficiency per kg. of gain	5.57	5.40	5.7
Dressing percentage	75.91	76.36	78.88
Final dressing percentage	56.59	67.41	68.31

2.3 SCOPE OF SWINE FARMING IN THE COUNTRY

Despite the fact that swine rearing is a profitable enterprise most of the progressive farmers even today hesitate to adopt it due to prejudice sentiments in our society against pig farming. For many others there is religious taboo against pig. A majority of meat eaters also do not consume pork. Inspite of these drawbacks, the consumption of pigs has greatly increased in recent years due to nutritional awareness in people. Pig is highly prolific breeder and cheap source of protein. Pork as source of animal protein is gaining popularity in India and pork products will be in greater demands in years to come.

The carcass yield of pigs has been known to be highest among food animals. The high carcass yield would attract the farming community because of the better monetary returns within the shortest possible time. High fertility in pigs gives an assured income to the rearer round the year. Improvement in the pork quality apart from increasing pork yield would make the pigs more economical to rear. To bring about improvement in the productivity of pork and other piggery products, extensive crossbreeding programme has been taken up to develop animals of large size, better feed conversion efficiency, high dressing percentage and quality pork. Consequently a number of piggeries have come up where hogs are bing raised on scientific lines and manufacture of piggery products is carried out under modern conditions. Such a programme if encouraged would help in controlling the rising prices of mutton/chevon etc. It is hoped that swine development will go a long way in solving protein hunger prevailing in our country.

2.4 MODERN OUTLOOK OF SWINE INDUSTRY

Efforts are on for improvement in pork production and pork quality by judicious crossbreeding programme to develop animals of larger little size, efficient feed conversion and high dressing percentage, which would make pigs more economical to rear.

2.5 PIG AS A MACHINE

Pig production is basically an enterprise of converting inedible or low quality food to the food rich in animal protein. Pig is said to be machine because it converts grain, pasture and other feed into pork used as human food. Pigs serves best as salvagers of by products, surpluses and such refuse that occur in production, processing and consumption of many kinds of foods.

2.6 PIGGERY - A COTTAGE INDUSTRY

A small unit of 10 piglets when reared for fattening will weigh about 600 to 800 kg. at market age and will fetch a sizeable profit, If a market is already created. *Creating a market in advance is essential for selling pork products. It is lack of market that makes this industry a failure.* In a large number of places this industry thrives well. It can function as a cottage industry engaging a large section of people. Piggery will promote self sufficient of the small and marginal farmers especially in backward areas.

2.7 PART OF S.F.D.A. PROGRAMME FOR UPLIFTMENT OF WEAKER SECTIONS

On the recommendation of the National Commission on Agriculture Govt. of India sponsored a special livestock programme from 1977-78 to improve the economic status of weaker sections of the society *viz.*, small/marginal farmers and agricultural labourers to involve them in increasing livestock productivity. This programme includes rearing crossbred heifers, poultry, sheep and pig production.

Uttar Pradesh is one of the largest states of country and has a large economically weaker population including scheduled castes and scheduled tribes. The government has introduced many programmes for their economic and social upliftment. The Department of Animal Husbandry has also devised several programmes for the weaker section of the society.

Pig production programme is being operated in four districts of Uttar Pradesh namely Meerut, Ghaziabad, Badaun and Partapgarh.

For establishment of piggery units of 3 to 4 sows and one boar subsidy is given @ 25 to 33.5 per cent but not more than Rs. 3,000 per beneficiary and 50 per cent in case of scheduled castes/scheduled tribes beneficiaries subject to a maximum of Rs. 5,000 each.

Implementing Agency

The Destrict Rural Development Agency (DRDA) has been established in every district, which is assisted by Assistant Project Office (A.H.) Veterinary Field Assistants and other office staff.

2.8 TAXONOMICAL CLASSIFICATION AND ORIGIN

Kingdom	-	Animalia
Phylum	-	Chordata
Subphylum	-	Vertebrata
Class	-	Mammalia

Sub Class	-	Utheria
Order	-	Ungulata
Sub Order	-	Artiodactyles
Family	-	Suidae
Genus	-	**Sus** Linn.
Species	-	**Vittatus, domesticus**
Sub Species	-	1. Scrofa 2. Philipinesis 3. Christatus 4. Strotzai

Scientific name of modern pig
Sus domesticus

Indian wild boar—**Sus scrofa** Cristatus Wagner] Found in foot of
Pigmy hog— **Sus salvanius** (hodgson)] Himalyas.

Origin and domestication

Period of domestication	*Area*
490 B.C.	China
800 B.C.	England

2.9 STATISTICS OF SWINE

Total population of pigs in India 18 millions (2001 estimates)

Table 2.3 Pig population of world and India (millions)

Year	1977	1982	1987	1992	2001*
India[1]	7.6	10.1	10.8	12.8	18
World[2]	666.3	763.8	839.9	855.8	923

*WWW.Fao.Org.

1. Agric. Research Data Book, 1998, Library Avenue, New Delhi-12
2. F.A.O. Production Year Book, 1977, 1982, 1987, 1992.

SWINE WEALTH :

As per 2001 livestock census, the total pig population in Inda is 18 millions and the country shares only 1.9 per cent of the world pig population. With an annual growth rate of 3.8 per cent, the pig number nearly doubled during past 20 years. The majority of the pig population of India is to be found in North Eastern sates, where nearly every family in the village roars pigs for meat. The crossbred and exotic pigs from only a small portion of the total pig population.

2.10 CHEMICAL COMPOSITION OF PORK

Moisture	77.4 per cent
Protein	18.7 per cent
Fat (Ether extract)	4.4 per cent
Minerals	1.0 per cent
Calcium	0.03 per cent
Phosphorus	0.20 per cent
Iron (mg.)	2.3 per cent
Calorific Value (per 100 g.)	114
Vitamin A (I.U. per 100 g.)	30.8
Vitamin B (I.U. per 100 g.)	60

Source : Health Bulletin No. 23 Nutritive value of Indian Foods and Planning of satisfactory diets (1938) Govt. of India.

2.11 IMPORTANT COUNTRIES IN HOG INDUSTRY (IN DESCENDING ORDER)

I. China
II. USA
III. Germany
IV. Russia
V. Poland
VI. France
VII. Denmark
VIII. U.K.
IX. Canada
X. Japan
XI. India

2.12. TERMS USED IN SWINE HUSBANDRY

Antinobacilloses : An animal disease mainly of cattle and pigs caused by bacteria **Actinobacillus lignieres.** It causes swelling and hardening of tongue and face. It is also known as wooden tongue.

Anorexia : Loss of appetite.

Anoxia : Deficient supply of oxygen to the tissues.

Back Fatter : A fat pig too heavy for bacon trade.

Baconer : A bacon pig.

Bacon : The meat from back and sides of a pig, preserved by curing with brine. Bacon may be sold "Green" or after wood-smoking strong flavoured "Smoked bacon".

Barrow : A male hog whose tastes are removed before reaching breeding age and before development of secondary sex characteristics.

Barren sow : A mature female pig incapable of producing offsprings due to infertility or sterility.

Blue Pig : Pig produced by crossing a white breed with black breed of pigs.

Boar : A male hog and sex organs intact and generally used for breeding (the uncastrated male of swine).

Brawn : A product prepared by boiling the meat of pig's head, trotters and tail.

Bristles : Stiff wiry hairs of the pigs obtained from back and neck parts.

Bull nose in swine (Infectious Rhinitis) : Diseased conditions of the pig snout due to infections one or more pathogenic micro-organisms. The process may be acute or it may run a chronic course resulting in facial distortion and meningitis, must not bé confused with Atrophic Rhinitis. The organism responsible for more cases is *Spherophorus necrophorus, Necrobacillus, Pseudomonas pyocyaneus : Brucella bronchiseptica or Akaligenes brochiseptica.*

Cad : The smallest pig of a litter usually last to be forwarded, also called as Rit, Crit, Criting or Ratling.

Carcass : The major portion of meat from body of pig after removal of Viscera, head, skin and shanks.

Chopper : Heavy built fleshy, pig, or tool (knife) used for reducing size of piece of meat by cutting.

Clean Pig : Castrated males and unserved gilts which are healthy and intended for slaughter.

Coupling : The act of breeding/mating in pigs.

Creep : An area of access of piglets separate from the sow lying area.

Creep fed : Given extra feed by means of small opening in panels that permit only smaller-sized animal (piglets, calves, lambs, kids).

Creep feed : The first solid and palatable feed provided to the young ones and is usually offered in creep. It is fortified with proteins, minerals, vitamins and antibiotics.

Creep feeding : A system of feeding young animals prior to weaning, it is designed to exclude mature animals.

Drove : A group or groups of pigs taken from place to place.

Fat Stock : Livestock fattened for sale in a market.

Fattening : Deposition of unused energy in the form of fat within the body tissue.

Feeder pigs : 2 to 5 months old, fast growing stage.

Finishing pigs/fattening pigs : The pigs weighing 45 kg and above, fed for slaughter purposes to produce carcass as desired finish.

Farrowing : Parturition /birth of piglets in sow.

Farrowing crate : A steel crate used in a pig pen to contain a sow and prevent it from lying on the killing the piglet, whilst allow them access to her treats.

Farrowing index : The number of litters produced by sow per year.

Farrowing rate : The number of sows actually farrowing as a percentage of sows served initially.

Farrowing interval : The average time for a sow or a group of sows between one farrowing and next.

Guard rails /farrowing rails : They are fitted at a short distance from walls and floor of pen. The loss of piglets from crushing may be reduced by use of pig brooders, farrowing crates or guard rails. It is usually the small less active pig that is crushed. A pipe or rail 20 cm to 25 cm above the floor and 20 to 25 cm. from the sides/walls of the farrowing pen will be sufficient. A 5 X 5 cm. satisfactory which can be installed.

Gilt/Hilt/Yelter : A young female pig kept for breeding purpose which has either not conceived as yet or going to farrow for the first time. She is also *hilt* or *yelter*.

Gilt replacement : A gilt introduced into breeding herd as a replacement for a cull sow.

Gravid sow/Gilt : Pregnant sow/gilt.

Growers/Growing pig : The piglets of either sex weighing from 10 or 12 kg to 45 kg live weight; Fast growing stage of 2 to 5 months.

Ham/Gammon : The thigh of a hog, prepared for a food. The hind leg of a swine from the hock upwards on the live animal.

OR

Salted meat from hind legs of pigs. It is also called Gammon.

Ham Shank : The hock end of a ham.

Hog : It is a general term used to denote the entire swine family or it is synonymous with swine.

OR

A pig above 55 kg body weight.

Hog Cholera : Acute, septicaemic, highly contagious and fatal disease of pigs, characterised by high fever, anorexia, severe leucopaenia, haemorrhages in different parts of the body. It is caused by virus.

Hog flue/swine fever : An acute, highly contagious and fatal disease of pigs. It is characterised by sudden onset with fever, respiratory symptoms and marked prostration. It is also caused by virus.

Hogg : An uncastrated male pig.

Hog Mange : Hog scab, or scabies is a contagious skin disease. It is caused by parasites *Sarcoptes scabiei Suis.*

Lard : The fat from the pig carcass after it has been tendered, *i.e.* melted down.

Litter : A group of piglets born to a sow/gilt in a single farrowing.

Loin : The part of quadrupt situated on both the sides of vertebral column between ribs and hip bone.

Maiden gilt : A female pig of 6 to 8 months of age.

Niddle teeth : Four pairs of small needle like brown coloured teeth found in newly born piglets, have no practical value. These may irritate the udder of sow. Therefore these are clipped off shortly after birth of piglets.

Packers : Old sows not suitable for consumption of fresh meat due to age or quality of meat. Pork from such sows is usually pickled, canned or cured into different forms.

Pen : A house for keeping a single pig or group of pigs.

Pig : "A young swine".

A domesticated animal "*Sus domesticus*" with long broad snout and body covered with bristles utp 55 kg. body weight.

Piglings, Piglets : "A small pig."

Baby pig upto 8 weeks of weaning age or 15 to 20 kg in weight.

Piglet anaemia : Blood condition caused by lack of haemoglobin due to deficiency of copper and iron, usually found in suckling pigs.

Piggery : A place where pigs are kept. A variety of pig housing are in use. Breeding pigs are sometimes run outside on well drained pastures and housed in movable sheds or temporary shelters. Store and fattening pigs are sometimes kept in covered dunging yards, usually covered with straw bedding. Modern in-door system particularly for farrowing pigs, usually consists of a low, insulated buildings provided with a heat source and controlled ventilation.

Pigging (Piggu) : Refers to a sow that has appearance of having recently "suckled' pigs or due to farrow soon.

Placenta : The membranes providing physiological link in the uterus between dam and offspring of facilitate the passage of nutrients and removal of waste products.

Pork : Fresh, frozen or salted meat with pig carcass in called pork.

Porkers : Pigs whose dead weight ranges between 27 to 36 kg.

Pre - starter : Those piglings that are being introduced to feed. They are 7 to 21 days old and weigh 2 to 5 kg (creep).

Pre weaner : Those piglets that are within 35 - 56 days old and weigh 11 to 23 kg.

Rig : A male pig/sheep with one undescended testis.

Rigor Mortis : The stiffness of body muscles observed shortly after the death of an animal. It is caused by an accumlation of metabolic products, especially lactic acid in the muscles.

Runt : A small undersized or stunted animal particulariy the smallest piglet in a litter known as runt also called ***Anthony, Cadme, Daniel, Dilling, Dolly Sharger or Wrenock.***

Sausage : Product prepared from fresh minced pork, free from bones and skin, usually having shoulder piece and meat trimmings cut from other portions of the carcass. It also includes 13 per cent fat trimmings or added along with wheat flour. This material is filled in intestines of pigs/sheep or insynthetic casings (membranes).

Service : The act of mating or copulation.

Shoat/shote : A young pig of either sex usually between 27 to 72 kg live weight.

Side : A butcher's term for a half of a carcass divided along the back bone.

Sow : Female pig kept for breeding purposes, which has farrowed at leas once or more number of times.

Sowbelly salt pork : Unsmoked fat bacon.

Spaying/soing : It is the removal of the ovaries by surgery to prevent breeding.

Stag : A male pig usually castrated after puberty.

Store pig : A pig between 8 to 15 weeks of age for market.

Stress : The sum of all non-specific biological phenomena caused by adverse influences. It includes physical, chemical and or emotional factors which causes physiological tensions and may contribute proneness to diseases.

Starter : Those piglings that have started to eat and are 21 to 35 days old. They should weigh between 5 to 11 kg.

Sty : House of pig/living place having one or more pens for pigs.

Suckling Pigs : Very young pigs which are being nursed by sow.

Suckling : Act of feeding milk by sow or her young piglets.

Swine : It includes all types of domestic pigs.

Swine erysipelas : An infestious bacterial disease caused by **Erysipelothrix rahasicpathiae** which is nonmotile, non capsulated Gram positive bacteria.

Swine fever : An infectious notifable disease caused by virus. It is characteristed by fever, anorexia, foul smelling diarrhoea, discharge from eyes, distressed breathing and general weakness. Death may occur in few days in its chronic form. It is also called Hog cholera/pig typhoid/ Hog flu.

Swine Paratyphoid : Infectious bacterial disease found in acute or chronic form characterised by fever, anorexia, red or purple discolouration of skin followed by diarrhoea. It is caused by **Salmonella choleraesuts**.

Swine Vasicular exanthema : An acute infectious disease of swine characterised by the formation of vesicles on the snout, mouth cavity, feet and other parts of body. It is also caused by virus.

Trottrs : The foot of a pig or sheep used as food.

Weaner : Piglet separated from mother for the purpose of independent rearing. These are two months old.

Weaning : Separation of young piglets from mother sow at 8 weeks of age.

Wallow : Water pool for pigs.

2.13 ADVANTAGES OF PIG RAISING IN INDIA

1. Efficient converter of concentrates into meat.
2. *Quick and net higher returns* because of 6 months of market age.
3. **Rapid expansion** of enterprise than cattle and sheep.
4. Relatively **less investment** on equipments, sheds, herd.
5. *More prolific* : Exotic breeds being more prolific may produce two litters per year and about 8 to 12 piglets per farrowing.
6. *Highest fat storing ability* as such no animal equals in this characteristics.
7. Pigs are *efficient convertor of many hyproducts* and feed into pork.
8. Feed conversion ratio is most efficient in pigs compared to other livestock (1 : 3 compared to beaf 1 : 5). They need less feed per kg gain of body weight.
9. Pigs require less roughages and hence a small acreage of pasture for growing and fattening pigs.
10. Requirement of labour in pig production is low because pigs are adapted to both self breeding as well as full feeding.
11. Fluctuations in prices of market hogs are less during the year.
12. Hogs have high dressing percentage (60-80%) and are usually in demand. Pig produces 65 to 80% of meat from feed consumed.
13. Hogs meat is well suited to curing and smoking.
14. Nonrecurring expenses are relatively less.
15. Pigs can very well utilize the kitchen garbage and left out feed articles of other livestock.
16. The hog enterprise can be started by anyone and at any time and can be closed at any time if found uneconomical.
17. Excels all other animals in fat storing ability.
18. Pig skins are used for light leather goods.
19. More meat can be produced from pigs per unit of time and cost.
20. Pork has high energy due to higher fat content and slightly lower water content.
21. Initial investment in setting up a piggery unit is very low.
22. Byproducts of pig industry such as skin, bristles, teeth and hoofs etc. are increasing in demand.

23. Pig manure is very useful and valuable for fields.
24. More flexibility and expansion ability of enterprise due to most prolific and quick growing animals.
25. Pork is relatively richer in phosphorus and iron content.
26. Source of animal protein.
27. Conversion of inedible feeds to edible food.

2.14 DEMERITS IN HOG RAISING

1. Utilization of larger percentage of concentrates.
2. Have direct competitions for cereals with human beings.
3. Pasture are spoiled and rendered unsuitable for other livestock.
4. More problem of parasitic infestation through pigs.
5. Control of diseases needs much time and labour.
6. Great variation in demands of pork in Indian market.
7. More labour at farrowing time.
8. Have religious taboo (in Muslims).

2.15 PURPOSES OF RAISING PIGS

(a) **Primary purpose** : Meat production
(b) **Secondary purpose** : Production of lard, pig, skin, bristles and manure.

2.16 MAJOR PIGGERIES

1. Andhra Pradesh Agriculture University, Tirupati, A.P.
2. Indian Veterinary Research Institute, Izatnagar, U.P.
3. Assam Agricultural University, Khanapara, Guwahati, Assam.
4. J.N.K.V.V. Jabalpur.
5. Keventer's Piggeries - Kolkata and Darjeeling.
6. British India Steamship Co., Kolkata
7. Bacon Factory, Ganavarum, A.P.
8. Allahabad Agricultural Institute, Deemed University, Allahabad.
9. The Piggery Unit, Govt. Livestock Farm, Hissar, Haryana.
10. The Govt. Piggery Farm, Ambala City, Haryana.
11. Ranchi Veterinary College, Birsa Agriculture University, Ranchi, Jharkhand.
12. Bacon Factory, Harringhata Farm, West Bengal.
13. Pig Breeding Farm, Pedony, Darjeeling.

3

Breeds and Breeding of Pigs

Pigs has 38 somatic chromosomes. There are about 60 breeds of domestic pigs in the world.

3.1 CHARACTERISTICS OF INDIAN WILD BOAR (Sus Scrofa Cristatus)

Colour - rusty grey in young, dark chestnut brown in adults, snout - long, ribs - short, ears-long, distinct sparse coat and full crest or mane of black bristles running from nape down to the back (Taylor *et. al* 1979). Male have both upper and lower tusks curving outwards from mouth. They live in groups of 10 - 20.

Pigmy hog (Sus salvanius)

They live in the dense forest along the base of Himalayan in Sikkim, Nepal, Bhutan, Assam, Colour is brown/black, ears-small, slightly long hair on hind part of neck and middle of back : have no distinct crest, Adult animals measures about 30 cm. and weights approximately 7.5 kg. They also live in groups.

3.2 INDIGENOUS DOMESTICATED PIG

Have no distinct breed features. Therefore, characteristics vary with topography any climatic conditions from region to region. This is black brown, grey, rusty-grey or blending/admixture of two or more of these colours. These are raised traditionally by weaker section of the community.

Swine Breeds

The desi pig (country hog) which has been evolved from gradual domestication of wild pig (*Sus scrofa cristatus*) are small-sized pigs with a body weight of 40 - 80 kg and are found in Uttar Pradesh, Bihar, Madhya Pradesh and Punjab sates of India.

Besides this, Ghori breeds of pigs has evolved from *Sus salvanius* hodgson is found in Manipur, Assam, Arunachal Pradesh, Nagaland, Mizoram, Sikkim and Tarai Himalayan region and are called as Dome pigs of pigmy pigs. Ankamali pigs evolved from domestication of *Sus scrofa* and *amanesis biyth* and are found in Kerala, Karnataka, Tamil Nadu and Maharashtra states of India.

Besides desi breeds, recognized exotic breeds *e.g.* large white Yorkshire, Landrace, Duroc and Hampshire are also found in India.

Recognising the importance of pigs in future, the State Government and Union Territories are running about 100 pig breeding farms throughout the country for upgradation of indigenous pigs with improved and exotic pigs.

These farms are rearing about 29,000 pigs of which 5,500 are exotic pigs of breeds like Large White Yorkshire and Hampshire. These farms breed pigs and supply piglets, boar and sows to the farmers for crossbreeding and improvement of desi stocks. The farms also run short courses and training programmes for pig rearers.

The Central Government has sponsored a scheme "Assistance of States for Integrated Piggery Development" for strengthening the infra-structure of pig breeding farms and marketing facilities of pig products in India. This scheme also includes the establishment of new pig breeding farms under Agricultural Universities and Krishi Vigyan Kendras.

3.3 EXOTIC/IMPROVED BREEDS OF PIGS

3.3.1 Classification of Swine breeds based on utility

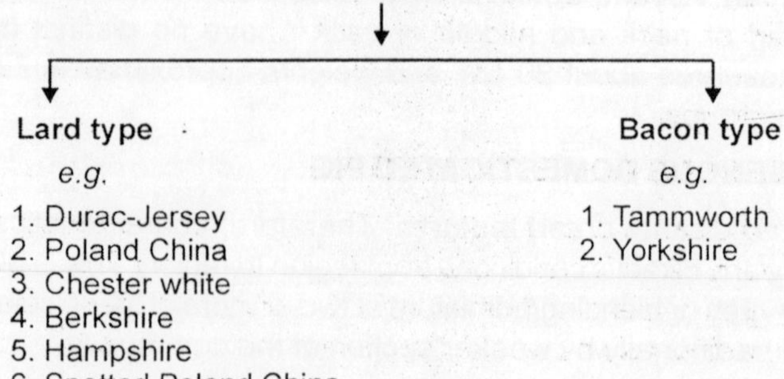

3.3.3 Exotic breeds experienced in India

1. Large white Yorkshire
2. Middle white Yorkshire
3. Saddle back
4. Tamworth
5. Berkshire

3.3.3. REGION-WISE CLASSIFICATION OF BREEDS OF PIGS

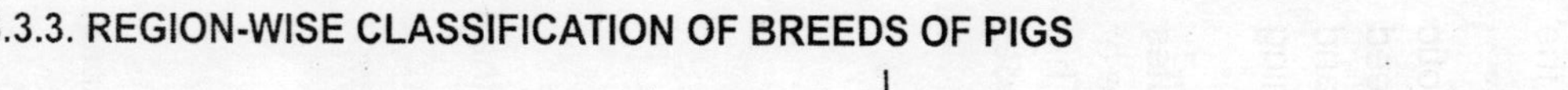

American breed	British breeds	Central and South American	African	Asian	Danish & other
e.g.	e.g.	e.g.	e.g.	e.g.	e.g.
1. Poland China	1. Berkshire	1. Cuino	1. Bakosi	1. Cantonese	1. Landrace
2. Durac Jersey	2. Large shite (Yorkshire)	2. Pelon		2. Malayan	2. Edelschwein
3. Chester white	3. Middle white	3. Cariollo		3. Balinese	
4. Hampshire	4. Tamworth	4. Nilo		4. Sumba	
5. Spotted Poland China	5. Large black	5. Canastra		5. Berkjala	
	6. Essex	6. Percira		6. Jalijala	
	7. Wessex Saddle back	7. Talu			
	8. Gloucester old spot	8. Pirapitanga			
	9. Welsh				
	10. Cumberland				
	11. National long white lop-eared				
	12. Large white Ulster.				

3.4 FACTORS EFFECTING SELECTION OF BREED

1. Availability of good breeding stock
2. Prolificacy
3. Growth ability
4. Temperament
5. Carcass quality
6. Efficient feed conversion
7. Nicking ability
8. Market demand and price
9. Disease resistance.

3.5 FACTORS FOR SELECTION OF BREEDING STOCK

1. Type and appearance
2. Pedigree
3. Health
4. Transmitting ability
5. Performance
6. Reliability of breeder
7. Size of litter (minimum eight piglets)
8. Strength and vigour of litters
9. Milking ability
10. Temperament
11. Gain and feed efficiency of progeny
12. Longevity
13. Fertility
14. Free from defects
15. Weaning weight of litter gilts = 120 kg; in sow = 150 kg.

Note :

1. Selection of individual animals is more important than breed.

2. There are more differences among individuals within a breed than between breeds.

Principle consideration for selection of breeding stock are as follows :

(A) **Type and appearance** : This includes -

1. General form, conformation and breed type : In general the pigs may be grouped into three categories -

(i) *Small and chuffy type* : They are lard type, short legged and compact in body quick maturing.

(ii) *Long legged and large type* : These are long bodied, high bcked, up standing and shallow in bodies.

(iii) *Intermediate type* : These are deep bodied, uniform in width and moderately long in body with large hams and legs of medium length.

2. **Size or weight for age** : Following average weight of pigs at different age may be taken as guideline :

Age	*Average Weight (Kg)*	*Age*	*Average Weight (Kg)*
1 to 7 days	1 - 1.5	5 months	50 - 60
7 - 21 days	2 - 5	6 months	60 - 70
21 - 35 days	5 - 10	7 months	80 - 90
57 - 90 days	30 - 35	10 months	160
90 - 120 days	40 - 45	12 months	180

3. **Development in the regions of high priced cuts** : High priced cuts are back and loins which provide pork chops, hind quarters which produce hams and sides give bacon. Therefore breeding pigs should not be fat but fairly fleshy having fairly wide backs and loins, plump rear quarters, long and deep sides.

4. **Feet and legs** : Legs should be straight, well apart and medium in length. Do not choose pig with defect in legs and feet as it may be transmitted to the offsprings.

5. **Quality** : It means certain amount of refinement such as medium sized bones, trim jowls, firm flesh, smooth shoulders, fine hair and skin, free from wrinkles etc.

6. **Teat development :** Sow or gilt should have good developed udder with at least 12 teats in number. Since teat character is heritable the gilt with blind teat or inverted or any other abnormality should not be selected. Such teats dont's produce milk and cannot nurse the young ones.

7. **Other heritable defects** : Selector should watch for swirls, whorls, or rosettes of hair as they are considered objectionable.

(A) Performance : A good to excellent performance for a gilt first litter is eight or more pigs raised to a weight of 120 kg and in case of sow not less than 150 at 8 weeks of age.

(B) Pedigree : Good pedigree record providing informations like birth, number animals in each of several generations.

(C) Transmitting ability/prepotency : A gilt or sow that transmits desired characters to her off springs in a steady manner is said to be highly prepotent.

(D) Health : The boar, gilt or sow selected should be healthy and immunised against all kinds of diseases.

(E) Reliability of breeder : The owner from whom breeding stock is being purchased should be truthful and reliable in providing informations on litter size and weight.

(F) Price and Market : One should make sure of getting good yield for the money spent. One should not compromise for the quality and must purchase a good and healthy stock.

3.6 CRITERION FOR IMPROVEMENT

1. **Litter size and livability** : The number of piglets born per litter and the number weaned out of them will determine profit of enterprise.

2. **Total litter weight at weaning** : It is a measure of the merit of pre-weaning performance in swine and is determined by the litter size, liability of piglets and growth rate. It is also an indirect measure of milking and mothering ability of the sow. To make selection easy, all pigs in a herd should be weaned at the same age.

3. **Economy of gain from weaning to market** : Daily gain from weaning to market influences profit because fast growing pigs have to be kept for less number of days thus requiring less labour, housing and care.

Weight at 154 days of age includes both pre and post weaning growth therefore an important economic trait. To standerdise pig weight to 154 days of age from weights taken earlier or later, adjustment factors have to be worked out as follows :

$$\text{Adjusted weight} = \frac{\text{Actual weight} + 154}{\text{Actual age} + 45} (199) - 154.$$

4. **Type and conformation** : These are important in determining potential production and carcass characters. Now a days, lean pigs are in demand. The aim is to produce pigs with the minimum fat and maximum primal cuts (ham, trimmed loin, shoulder and bacon) with desirable qualities of tenderness, juiciness and aroma. Rapid changes can be brought about in the conformation of pigs. Pigs carcasses with good conformation tends to be better than poorer conformation.

5. **Carcass traits** : Consumers demand is for more lean and less fat in pork. Quantitative carcass traits includes dressed weight, dressing percentage,

3.7. CHARACTERISTICS OF IMPORTANT BREEDS OF PIGS EXPERIENCED IN INDIA

Breed	Native Place	Head and Neck	Body Features	Mature Body weight
1. Large white Yorkshire	English	Longer, face dished broad snout, erect ears, long thin and forward head.	Entire-white, black, pigment spots long levelled back, fine skin, without wrinkles and smooth Prolific breeder.	Boar-300 to 400 kg. Sow - 230 to 320 kg. Principally bacon type.
2. Middle white Yorkshire	England	Short head and neck, jaw straight and light, unturned dished face	Cross between large and small Yorkshire, long levelled back white in colour, fine skin, medium sized, smooth skin, small bones.	Boar-249 to 340 kg. Sow-181 to 272 kg. Less prolific than large white, bacon type.
3. Landrace	Switzerland	Longer snout, dropping ears, medium sized, straight face and of medium length	Short legs,, lean carcass, white in colour with black skin spots, longer body, skin without wrinkles.	Boar-300 to 350 kg. Sow-200 to 250 kg. Bacon type excellent.
4. Berkshire	England	Short unturned nose, dished face, erect ears, inclined forward.	Medium sized, legs of medium length, black in colour, good width and broad back, hams and shoulders are well fleshed.	Boar 272 to 223 kg. Sow-204 to 294 kg. Meat is good quality.
5. Hampshire	U.S.A.	Head and Tail are black, erect ears.	Black hog, with a belt encircling the body including front legs, short legs, smaller breed.	Boar-250 to 320 kg. Sow-200 to 300 kg. Meat type midway between bacon and lard type.
6. Tamworth	U.K.	Longer head, long snout.	Golden brown in colour, thin shoulders, Strong back.	Boar-200 to 320 kg. Sow-180 to 320 kg.
7. Wessex Saddle Back	England	Longer head, straight, snout, ears forward.	Robust make up, black in colour, bacon type, prolific.	Bacon best quality. Boar-240 to 280 kg. Sow-180 to 260 kg.

carcass length, back fat thickness loin eye area, primal cuts percentage, meat to bone ratio. The other carcass traits qualitative in nature are sheer force value (kg/cm), fibre diameter (micron), pH & water holding capcity etc. The external appearance is only a very inadequate indicator of carcass quality.

3.8 ECONOMIC TRAITS OF SWINE

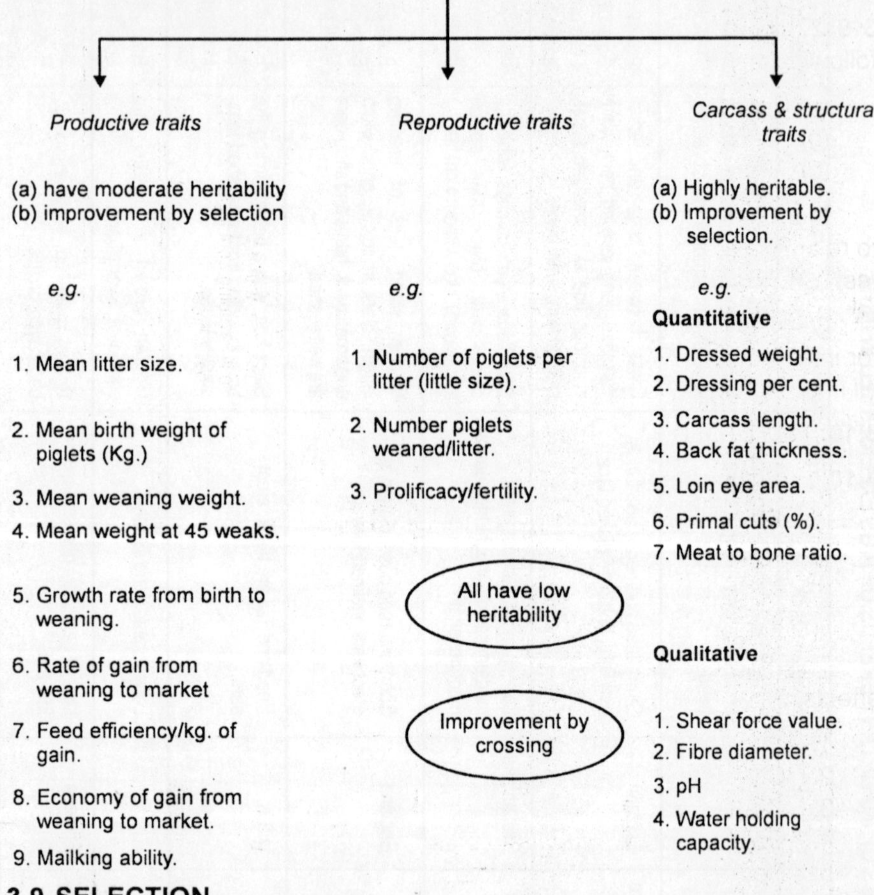

3.9 SELECTION

In includes choosing the parents of next generation. Selection responses to a particular trait depends upon the selection differential and heritability of trait. The selection differential in standard deviation units is called intensity of selection.

3.9.1 Parameters for Appraising Growth Potential

1. Average daily gain during specific post weaning period.
2. An index combining-average daily gain, post weaning and weaning weight.
3. Measurement of weight for age.

3.9.2. Methods of Improvement : This depending upon the following :

(a) Heritability of the characters.
(b) Whether trait is sex-linked/limited.
(c) Whether trait is measurable.

More direct the selection is, cheaper it will be. Emphasis must be given to relative cost efficiency of improvement by performance testing, progeny testing and nucleus testing.

Of these nucleus method is considerably cheaper and gives flexibility for introducing new genetic material and hence this is likely to be basis of all improvment schemes.

3.10. BREEDING SYSTEMS

3.10.1 In breeding

It involves mating of related animals for following purposes :
(a) To increase homozygosity in progeny.
(b) To develop inbred lines.
(c) To keep animals pure bred.

This system is not adopted commercially for the following adverse effects :

1. Decreases mean litter size with increase in age of pigs.
2. Causes slight decrease in post weaning weight.
3. Causes decline in milking and mothering ability of sows.
4. Delays sexual maturity in pigs.
5. Decreases sex libido in boars.
6. Reduce fecundity and prolificacy.

3.10.2 Out Breeding

It consists of mating of unrelated animals and systems of it are being extensively used for achievement of good results with regards to performance of pigs.

3.10.2.1. Out Crossing

This is the common method of breeding and multiplying pure bred swine which involves mating of unrelated animals of the same breed. Compared to inbreeding this system is better for crossing inbreds for following advantages :

1. Keeps animals pure bred.
2. Slight gain in litter size.
3. Helps to regain in vigour in animals.
4. Some increase in livability and growth rate.

Limitations

1. Poor performance of inbreds.
2. Higher cost involved in development of inbred lines.

3.10.2.2. Grading-up

The non descript indigenous pigs forms the bulks of the pig population in the country. It involves use of boars of improved breeds with indigenous hogs. It will be advantageous to grade up bulk of indigenous hogs by successive use of boars of large white Yorkshire or Landrace breed. These two breeds were used in All India Coordinated Research Projects in the country. Following are the merits of this system :

1. Causes Improvement in productive traits of vast population of non-descript pigs in India.
2. Suited to areas where high quality pure breeds cannot be maintained due to poor management and feeding conditions.
3. Increases fertility and prolificacy in successive stages.

3.10.2.3. Cross-breeding

For commercial swine production programme it is common method used in areas around bacon factories. It involves mating animals of two different breeds, *i.e.* crossing of Landrace sow with Yorkshire boar or *vice-versa*. Following are the merits of this method :

1. Fewer embryonic losses.
2. Causes increase in litter size and uniform birth weights and weaning weight.
3. Crossbred sow wean larger litters/more weight at weaning.
4. Greater resistance to environmental stress.
5. Increase in growth rate.

6. Early age of maturity.
7. Increases livability of pigs and high vigour.
8. Regularity in breeding.
9. Increased efficiency of feed conversion.
10. Mothering ability and higher milk production.

Note : Performance of crossbreds compared to purebred middle white Yorkshire and nondescript (Desi) Pig is presented in Table 3.1 and 3.2.

Table 3.1. Performance of indigenous, exotic breed and crosses in India.

Traits	*Desi*	*Middle white Yorkshire*	*Crossbred*
Percentage of piglet born dead.	7.68 ± 0.61	7.84 ± 0.81	3.87 ± 0.43
Mean litter size	7.79 ± 0.02	8.71 ± 0.83	9.53 ± 0.61
Mean birth weght of piglet (kg.)	0.90 ± 0.02	1.34 ± 0.03	1.23 ± 0.04
Mean weaning weight (kg.)	4.02 ± 0.23	7.01 ± 0.30	6.65 ± 0.36
Mean weight at 48 week (kg.)	40.0 ± 0.90	73.5 ± 0.95	52.0 ± 0.84
Growth rate from birth to weaning (kg.)	0.073	0.108	0.102
Feed efficiency/kg. of gain	5.57	3.40	4.70
Dressing percentage	75.91	76.36	78.88
Final dressing percentage	56.59	67.41	68.31

Table 3.2. Growth and feed efficiency of indigenous and large white Yorkshire Pigs.

Parameter	*Indigenous variety*	*Large white Yorkshire*
Initial body weight	28.00 ± 1.62	14.50 ± 0.49
Final body weight at slaughter (kg.)	59.17 ± 0.51	63.08 ± 1.41
(A) **Grower period** (upto 35 kg. live weight)		
Number of pigs	13	22
Period of reach 35 kg. weight	53 ± 7 days	48 ± 3 days
Feed consumed (kg./day)	1.28 ± 0.10	1.35 ± 0.03
Daily gain (g)	223 ± 17	458 ± 17
Feed per kg. gain (kg.)	5.74	2.94
(B) **Finisher period** (upto slaughter)		
Number of pigs	19	22
Period of reach slaughter weight (days)	102 ± 7	79 ± 2
Feed consumed (kg./day)	1.74 ± 0.04	1.38 ± 0.08
Daily gain (g)	250 ± 18	363 ± 19
Feed per kg. gain (kg.)/FCR	6.93	3.92

Parameter	*Indigenous variety*	*Large white Yorkshire*
(C) **From initial weight to slaughter**		
Number of pigs	19	22
Period of reach slaughter weight	138 ± 11 days	126 ± 2 days
Feed consumed (kg./day)	1.65 ± 0.04	1.36 ± 14
Daily gain (g)	2.44 ± 15	1.36 ± 14
Feed per kg. gain (kg.)/FCR	6.74	3.52
Carcass characters - Quantitiative :		
Dressed weight (Kg.)	47.18 ± 0.44	49.66 ± 1.30
Dressing percentage	80.06 ± 0.64	78.62 ± 0.72
Carcass length (cm)	62.63 ± 0.61	78.96 ± 0.55
Back fat thickness (cm)	3.02 ± 0.14	2.33 ± 0.10
Loin eye area (cm^2)	18.73 ± 0.54	27.43 ± 0.78
Primal cuts (%)	65.85 ± 6.47	60.12 ± 0.64
Meat to bone ratio	3.38	4.45
Carcass characters - Qualitative		
Sheet force value (lb/sq")	9.45 ± 0.14	5.97 ± 0.14
Fibre diameter (micron)	56.68 ± 0.64	45.24 ± 0.85
pH	5.59 ± 0.02	5.62 ± 0.01
Water holding capacity (cm^2)	1.40 ± 0.5	2.92 ± 0.09
Proximate composition of L dorsi muscle		
Moisture %	71.52 ± 0.27	69.36 ± 0.55
Crude protein %	22.27 ± 0.24	24.64 ± 0.25
Ether extract %	4.58 ± 0.15	2.66 ± 0.09
Total ash %	1.58 ± 0.09	2.11 ± 0.09

3.11 STARTING A PIG FARM

For a medium size farm a unit of 10 sows and one boar will be enough. For a beginner a small unit of 5 sows and 1 boar will be advisable. For initial lower cost it will be better if started with young piglets.

3.12. SELECTION OF BOAR : Consider following parameters :

1. True to the breed.
2. Masculine appearance
3. Long deep body.
4. Strong legs and smooth shoulders.
5. Sound health and performance record.
6. No cryptorchid condition.
7. Age between 1.5 to 2 years.

8. Select only fertile boar with well developed testes.
9. To overfat conditon.
10. Strong back.
11. Active look

Note :

1. A irritable boar difficult to drive and one who incliens to fight may transmit a nervous disposition to piglets. This may make them poor mothers.

2. The litter size has been found to vary significantly (P < 0.01) between boars but not between seasons. This indicates the necessity for selection of boars based on the litter size to obtain maximum offsprings.

3.13. SELECTION OF SOWS : Consider following parameters in view :

1. Sow must be from a litter whose litter size and weight at birth and weaning weight is maximum.
2. Have a minimum back fat thickness.
3. Sow must have well developed udder with twelve teats and at least 6 teats in each row, evenitly distributed on belly sides.
4. Teat of sow must be free from any abnormality.
5. Sow must have deep body.
6. Select the sows that are already bred at least once.
7. Age of brood sows must be 2 to 3 years.
8. Sow must produce numerous young ones each year.
9. Sow must have good mothering ability.
10. Sows must be ready for rebreeding at the end of lactation.
11. Select gilt/sows that are healthy.
12. Sows of offsprings must thrive well.
13. Sow must have quiet disposition.

Note : Many farmers make errors of depending much on boar quality because boar is considered to be "half the herd" in the matter of inheritance, but in swine there are so many youngs. Motherliness of sow can make difference in raising 5 to 8 piglets out of 10 farrowed.

2.14 CULLING OF ANIMALS

Boars

1. Infertile ones.
2. Boars of over 5 years age.
3. Irritable nature and nervous disposition.
4. Over fat and too heavy, finds difficulty to mount.

5. Not true to the breed.
6. Cryptochid.
7. Weak limbs.

Sows

1. One third of older sows annually.
2. Gilts or sows not settled after 4 days breeding period.
3. Nervous and irritable nature.
4. Produce small litter.
5. Sows with defective teats and poor milkers.
6. Sows with small vulva is an indication of internal reproductive defects.
7. Sows/gilts with inverted teats.
8. Gilts and sows which do not meet the standard of meaty hogs.

3.15. GUIDELINES FOR NORMAL REPRODUCTION OF PIGS

Age at pubery	6 to 7 months
Breeding age of gilts	10 to 12 months
Breeding weight of gilts	90 to 100 kg
Breeding age of boar	18 to 24 months
Number of sows per boar	10
Heat cycle	19-23 days (Ave. 21)
Heat period	2-3 days
Mating time	Gilts-First day of heat and in sows second day of onset of heat.
Number of services per conception	Two, at interval of 12 hours.
Gestation period	112-114 days
Sucking period	50-60 days
Average litter size at birth	10 to 14
Average litter size at weaning	8 to 10
Rest period	45 days
Occurrence of heat after weaning	2 to 10 days
Period of mating	15 days after weaning
Volume of ejaculate	200 c.c.
Average number of sperms/cumm.	100,000
Average age to castrate pigs	4 to 8 weeks
Market age of fattening pigs	6 months
Farrowing interval	7 to 7½ month
Years-sows are known to breed	8 to 10 years
Average life of sow	6 litters

Note : Boars with low protein intake have reduced libido. This reduction in libido may be a result of decreased estrogen concentration in circulation.

3.16. MANAGEMENT AT BREEDING OF PIGS

Feeding gilts and dry sows liberally to increase energy intake 10 to 15 days prior to mating is called flushing. It may be done as follows :

(a) Feed leguminous hay (Cowpea/lucerne/berseem) for it supplies more protein, minerals and vitamins).

(b) Extra allowance of grains.

(c) Give multivitamin injection along with "Flushing".

Note :

1. Boars fed low level to protein take longer to mount a dummy and start ejaculating and produce less sperms.
2. Pandey and Singh reported highly significant correlation coefficient between farrowing percentage and mean percentage of seminal characteristics of preserved boar semen.

3.17 FLUSHING

Feeding extra allowance of grains by about 0.5 to 0.7 kg before breeding season is called flushing. This is done to increase energy level in diet for increasing ovulation rate in gilts and sows, so that they may gain weight from 0.5 to 0.7 kg daily from 1 to 2 weeks before breeding season.

Advantages of flushing

1. Improvement in physical condition of female.
2. Prompt post wearing oestrus.
3. Shows prominent heat symptoms.
4. Increases ovulation rate
5. Good litter size.
6. Shortens period between weaning to successful conception.
7. High number born.
8. More uniform litter size
9. Minimize embroyonic losses.

3.18. DETECTION OF "HEAT" IN SOWS

Following are the symptoms of heat in sows :

1. Vulvar swelling and redness.
2. Vaginal discharge.

3. Urination frequently.
4. Reduced appetite.
5. Mounting behaviour.
6. Immobility when normal manual pressure is applied on the back region (Limbo-Sacral) "Standing heat".
7. Restlessness and excitement.
8. Mucus discharge from vulva.
9. Peculiar grunting sound.
10. Erection of ears when pressure applied on back.

3.19. OPTIMUM TIME TO BREED SOWS/GILTS

Standing heat as detected by immobility of sows in oestrus particularly exhibited by erection of ears when manual pressure is applied on the back of sow.

3.20. INFLUENCE OF BOAR CONTACT ON AGE AT PUBERTY IN GILTS

Five minutes of daily contact with mature boar is sufficient to stimulate early puberty in gilts providing gilts have adequate opportunity for physical contact with the boars. Gilts that are of 165 days of age appear to require daily boar exposure to obtain rapid and maximum pubertal response.

3.21. STILL-BIRTHS

Shobamani in a study reported that age of dam, environmental temperature during farrowing season and aflatoxicosis had badly influenced the rate of still births. Sex of the piglets had no significant effect. Main cause of still births reported by Sharpe and Lai *et. al.* were congenital abnormalities, mechanical damage to the piglet during birth and infections diseases.

Sows with history of regular still birth should be culled from breeding stock. Apparently stillborn piglets can be revived by cleaning upper respiratory tract promptly.

3.22. PREGNANCY DIAGNOSIS

Dyck (1992) compared all the prevailing methods of pregnancy diagnosis in the sows and observed that vaginal biopsy technique had many advantages *viz.*, high accuracy and low cost over other methods. Such as ultra sound and hormonal assay. Singh and Sinha observed the pregnancy in sows by Vaginal Biopsy Technique and reported that there were 2 to 3 parallel layers of vaginal epithelium with dark stained nuclei. The pregnancy diagnosis was found to be 96 per cent accurate. Diehl and Day also found an accuracy rate of 95.5%.

4

Feeds and Feeding of Pigs

4.1 DIGESTION OF FOOD IN PIG (FIG. 4.1)

Pig is an omnivorous takes feed to both animal and plants origin. Pig being non-ruminant cannot utilise fibrous fodders efficiently. Due to its simple stomach ration must have more of concentrates preferably in ground form.

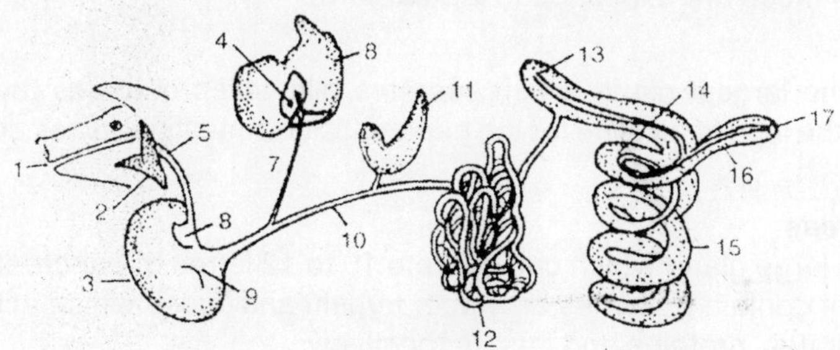

Fig. 4.1. Digestive organs of pig.

1 - mouth	2 - Parotid gland	3 - Stomach duct
4 - Gall bladder	5 - Oesophagus	6 - Oesophagus region
7 - Bile	8 - Liver	9 - Pyloric gland region
10 - Duodenum	11 - Pancreas	12 - Jejunum
13 - Caecum	14 - Colon	15 - Coils of colon
16 - Rectum	17 - Anus	

Digestive System of Swine (Fig. 4.1.)

1. Mouth
2. Oesophagus
3. Stomach
4. Liver
5. Pancreas
6. Small intestine
7. Caecum
8. Large Intestine
9. Rectum

1. Mouth

Food in mouth is broken into smaller particles by chewing, and mixed with saliva to make it into paste and swallowed. Three sets of salivary glands in mouth are :

(*a*) Parolid - below the ear behind the lower jaw,
(*b*) Submaxillary - below the lower jaw, and
(*c*) Sublingual - below the tongue.

The saliva in swine contains amylase enzyme which acts on starch and glycogen and converts into dextrins, maltose and glucose. The food is then pushed to pharynx and down to through oesophagus to stomach.

2. Stomach

Food in stomach is churned up by the squeezing action of the muscular walls and the gastric juice is added to it. Gastric juice produced by gastric glands in the walls of stomach contains mucus, hydrochloric acid and pepsin. Actions of these are explained in tubular form.

3. Liver

It is the largest gland which secretes bile which changes the acidic nature of the food to alkaline for the pancreatic and intestinal juices enzymes to act.

4. Pancreas

It is a large gland which can secrete 10 to 12 letres of pancreatic juice in a day. It contain enzymes amylase, trypsin and lipase which acts upon carbohydrates, proteins and fats, respectively.

5. Small Intestine

Juice from pancreas, liver and small intestine completes the process of digestion in the small intestine. Intestinal juice contains several enzymes and their actions have been explained in tubular form.

6. Caecum

The beginning of large intestine get enlarged to form a blind fut called caecum which harbours large number of bacteria which help in digestion of crude fibre.

7. Large Intestine

All the digested food moves to large intestine where water and other nutrients are absorbed from the digested food.

8. Rectum :

It is the last part of digestive system through digested and unabsorbed material called faecal matter is secreted out of anus.

DIGESTIVE PROCESS IN PIGS

Digestive Part	*Source of Secretion*	*Stimulus*	*Digestive fluids*	*Enzyme*	*Action*
I. *Mouth*	Salivary galns 1. Paroitd Submaxillary Sublingulal	Psychic, Mechanical, chemical	Saliva	Ptyalin (Amylase)	Glycogen Starch Dextrin Maltose and oligosaccharides
II. *Stomach*	Gastric glands 1. Parietal cells—HCl 2. Neck cells secrete mucin 3. Chief cells secrete Pepsinogen	Reflex, Gastrin Enterogastrine (inhibitory)	Gastric juice (Acid—HCl, Mucin, Intrinsic factor) Secreted from stomach glands of pigs.	Renin Pepsin No lipases	1. Coagulates soluble-protein of milk. 2. Protein -Peptone. 3. Tributyrin Fe+++HCl to Fe++ 5. Swelling to protein and anti bacterial effects. 6. Mucin-acid combining power to mucosa. 7. Intrinsic factor for absorption of cobalamin.
III. *Small Intestine*	Gall bladder	Choiccystokinine	Bile juice pH = 7.8		Emulsifies fats, stabilise emulsion Increases action of pancreatic lipase. Helps in excretion of pigments and cholesterol.
Three parts :	Pancreas	Scretin pancreozymin	Pancreatic juice (Secreted	Trypsinogen	Activated to trypsin by enterokinase.

1. *Duodi num*			8 to 9 litres per day in pigs,	Trypsin	Hydrolyse peptides bonds.
				Chymotrypsin	-- do--
2. *Jejunum*	Brunner glands in duodinal mucosa		Duodinal juice PH = 8.6	Carboxypeptidase	Hydrolyse peptides linkage with free —COOH
				Polynuclotidase	RNA and DNA to mononucletides,
				Amylase	Starch destrins and meltose to hexoses.
				Lipase	Fats to monoglycerides and fatty acids and glyccrol
3. *Ilium*		Enterokinin	Intestinal Juice	Aminopeptidase	Amino acid NH2 group
				Dipeptidase	Dipeptide to Amino acid
				Erepsion	Peptones to Amino acid
				Nuclcosidase	Nucleosides to Purins, Pyrimidine and pentose
				Nuclocotidase	Nuclcotides to nucleosides and H_3PO_4
				Alkaline-Phosphatase	Organic phosphate to free phosphate etc.
				Disaccharidase	Mahose, lactose, sucrose to hexoses
				Invertase	Sucrose-Invert sugar (Glucose + fructose)
				Lactase	Lactose to Glucose and Galactose
				Maltase	Maltose to Glucose
				Sucrose	Sucrose to Glucose and fructose.

Mode of Intake of Food

Large particles of ration and also fodder grasses are thoroughly crushed, chewed and masticated with the help of molar teeth. When ground concentrates are fed after mixing with butter milk/skim milk or water, pig dips its snout to the bottom of feed through and sucks the feed with the help of tongue. The part of the mouth will remain above the surface because angle of the jaw is situated far back. This part of the mouth above the surface of level of feed, sucks air together with mixed feed. This action of mouth along with rapid chewing movement with open lips gives a typical characteristic surping and smacking sound, found in uptake of feed in pigs.

Digestive process : It is briefly explained in the tubular form.

Digestion in Caecum and Colon

Pig has a large caecum and colon which indicates that pig is more related to the herbivores than to the carnivores. Owing to large caecum and colon pigs can digest crude fibres to some extent (6-7%) which depends upon bacterial fermentation in these organs. There are 10^8 to 10^9 micro-organism per gm of caecal contents which are mostly *streptococci* and *bacilli.*

The bacterial fermentation of cellulose leads to production of volatile fatty acids (acetic acid - 62 per cent, propeonic acid - 28 per cent and butyricacid - 18 percent) These VFA are rapidly absorbed in the blood system and utilized in the body. Compared to caecum, the colon is the site of greatest microbial activity which is chiefly responsible for crude fibre digestion. Micro-organism in these parts are capable of synthesizing several water soluble vitamins necessary for the body. Thus a normal supply of these essential elements to body is regulated.

When the food nutrients are completely digested they are absorbed into the body and undigested fraction of feed is excreted as faeces from the anus.

UTILIZATION OF FOODSTUFFS

Products of Digestion	*Intermediate Products*	*End Products*
	Proteins	
Proteoses	Tissues proteins	Carbon dioxide
Polypeptides	Blood sugar and liver glycogen	Urea, ammonium salts
Peptides	Tissue fat	Creatine, creatinine, and uric acid
Amino acids	Hormones (*i.e.* thyroxin epinephrine, etc.)	Sulfates and phosphates

Products of Digestion	*Intermediate Products*	*End Products*
	Alpha-keto and hydroxy acids Pruines, creatinine	Amino acids and proteins Water
	Carbohydrates	
Glucose (blood sugar)	Hexosephosphates	Carbon dioxide
Fructose	Triosephosphates	Water
Galactose	Pyruvic Acid	Lactic acid
Dextrins	Acetyl coenzyme A	
Maltose	Glycogen	
Pentoses	Tissue fats	
	Fats	
Fatty acids	Depot of fats.	Carbon dioxide
Glycerol	Phspholipids and cerebrosides	Water
Emulsions (with bile acids)	β-keto acids	Acetone bodies
	Acetyl coenzyme A	Synthesis of glycogen fatty acids, cholesterol, proto-pophyrin, uric acid and dicarboxylic amino acids.
	Minerals	
Soluble salts	Soluble salts Organo-metallic compounds	Soluble salts, eliminated by kidneys.
	Bone and teeth structures	Insoluble salts by intestines.
	Vitamins	
Soluble organic compounds	Enzymes and coenzymes in cellular reactions, especially those involved in oxidation.	In part eliminated by kidneys intestines, Partial oxidation and elimination as derived products.

4.2 SOME CHARACTERISTIC FEATURES OF SWINE NUTRITION

1. Pigs have a single compartment of simple stomach.
2. Being non-ruminant they cannot utilise fibrous fodder efficiently.
3. Ration must have more concentrates and less roughages.
4. Feed costs about 75 per cent of total cost of rearing.
5. Nutrition affects growth, production and reproduction.
6. Pig is an efficient converter of concentrate into meat
7. Feed conversion ratio is most efficent (1 : 3) and hence needs less feed per kg. gain.
8. Pigs can very well utilize the kitchen garbage and left out feed articles of other livestock.
9. Converts inedible feeds to edible meat.
10. Pigs directly complete with human beings for cereals.

11. The energy requirements are usually expressed as the amounts of DE/ME per kg. of daily ration.
12. Energy plays an important role for meat type of hogs.
13. Pigs suffer more from nutritional deficiencies than do ruminants.
14. Feeding programme must be efficient to make it profitable.
15. Always give free excess to water.
16. Three types of rations are fed before they reach maturity, *i.e.* creep/starter, grower and finisher.
17. Most economic feed ingredients must be selected.
18. Small pasture is good for raising pigs provided it has succulant green forages.

4.3 PARAMETERS AFFECTED BY QUALITY OF RATION

1. Growth rate
2. General resistance to diseases and parasiticism
3. Regularity in breeding.
4. Size of litter.
5. Vigour of litter.
6. Amount and quality of milk.
7. Carcass quality.
8. Dressing percentage.

4.4 FEEDS FOR HOGS

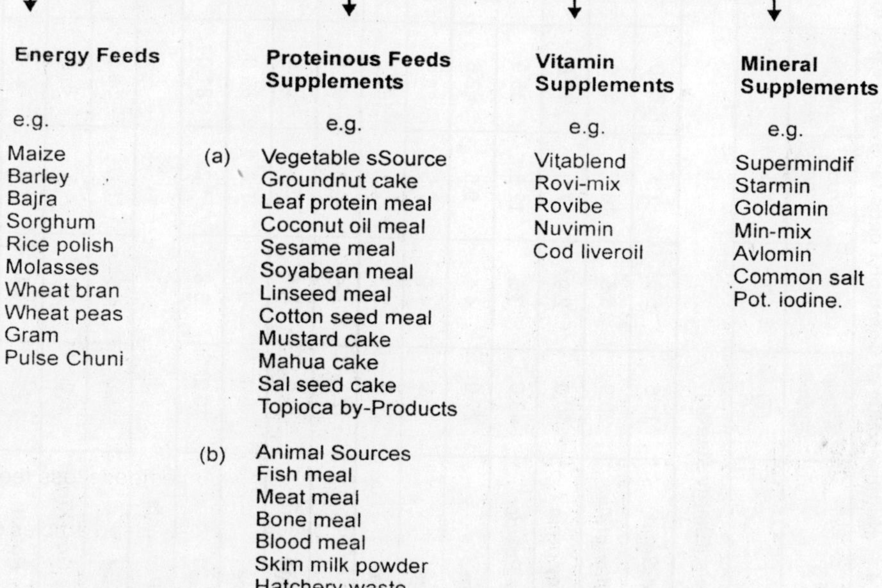

Miscellaneous feeds : Brewers' and distillers grain, papaya, Banana, Citrus fruits, Pumpkin, Pasture grass, Sweet Potato, etc.

Table 4.1. Nutritive value of Common Feeds for Swine (Value in Percentage on Dry Matter Basis).

Feeds	Crude Protein	Ether Extract	Crude Fibre	Nitrogen free extract	Ash	DE* Meal/kg.	ME* M Cal/kg.	Lysine %	Methionine %	Tryptophan %
(1)	(2)	(3)	(4)	(5)	(6)	(7)	(8)	(9)	(10)	(11)
I. Cereals (Energy source)										
1. Bajra Grain	11.5	5.0	1.5	79.0	3.0	2.64	2.35	0.38	0.28	0.19
2. Distiller grain	20.0	5.5	16.5	54.0	4.0	3.70	3.30			
3. Barley grain	10.0	2.9	7.0	77.6	2.5	3.75	3.33	0.5	0.5	0.32
4. Guar Powder	43.0	3.5	12.5	35.0	6.0	3.83	3.38			
5. Gram	20.0	4.1	9.8	62.5	3.6	3.71	3.30			
6. Maize	9.0	4.0	2.3	83.5	1.2	5.18	4.77	0.24	0.91	0.45
7. Jowar	12.6	3.3	22.1	80.0	2.0	3.57	3.17	0.27	0.10	0.12
8. Oat	12.5	4.5	13.0	66.5	3.5	3.44	3.07	0.43	0.15	0.55
9. Wheat	10.5	2.5	2.0	82.5	2.5	3.99	3.55	0.42	0.22	-
10. Rice	8.4	0.9	0.4	89.0	1.3	5.15	4.80	0.27	0.27	0.09
11. Ragi	10.2	3.8	1.2	81.0	3.8	3.88	3.50			
II. By-Products										
1. Wheat	15.0	4.5	15.0	59.5	7.0	3.1	2.76	0.59	0.21	0.18
2. Rice Grit (Kinki)	8.0	0.5	1.5	87.8	2.2	3.0	2.42	0.50	—	0.10
3. Rice Polish	13.0	15.0	3.0	60.0	9.0	3.25	2.66	0.50	—	0.10

4. Arhar Husk	6.0	2.0	41.5	45.5	5.0	2.40	1.98			
5. Banana Stem	2.5	3.3	20.5	60.5	13.2	-	-			
6. Gram Husk	5.0	0.6	49.0	39.0	6.4	2.59	2.31			
7. Maize cobs	2.0	0.8	36.4	58.0	2.8	-	-			
8. Pulse husk	11.2	1.5	26.5	47.8	13.0	2.43	1.99			
9. Pulse bran/chunni	12.5	3.5	25.5	51.4	5.6	2.60	2.12			
10. Skim milk Powder	36.9	0.2	-	54.1	8.8	-	-	2.45	0.85	1.85
11. Cow pea husk	7.5	4.5	24.5	51.2	12.3	2.45	2.0			
12. Moalsses	2.3	-	-	78.2	10.5	-	-			
III. Cakes (Vegetable Sources of Protein)										
1. Cotton seed cake	27.5	9.5	18.5	38.4	6.3	3.64	3.26	1.70	0.65	0.65
2. Mustard cake	33.0	6.5	11.5	42.5	6.5	3.95	3.53	1.29	0.38	0.31
3. Ground nut cake	46.0	6.5	6.5	33.5	7.5	4.05	3.58	1.30	0.44	
4. Linseed cake	38.0	5.0	10.5	33.5	6.5	3.44	3.06	1.21	0.68	1.75
5. Mahua cake	18.0	17.0	5.7	20.3	9.0	2.64	2.16			
6. Rape cake	35.0	10.0	8.0	38.0	9.0	4.14	3.71			
7. Sunflower cake	45.0	3.0	18.0	25.8	8.2	3.10	2.55			
8. Sesame cake	48.0	5.5	8.5	28.0	10.0	4.33	3.84	1.27	1.22	2.06
9. Soyabeen cake	48.5	5.0	7.0	33.0	6.5	3.72	2.68	2.86	0.60	0.28
10. Coconut cake	22.5	7.0	13.0	50.0	7.0	3.04	2.7	0.52	0.25	0.61

(1)	(2)	(3)	(4)	(5)	(6)	(7)	(8)	(10)	(11)	(12)
IV Animal byproducts										
1. Blood meal	88.0	1.8	1.0	3.0	6.2	-	-	7.0	1.14	3.0
2. Bone meal	54.0	10.0	2.2	2.8	31.0					
3. Fish meal	68.0	7.5	1.0	3.5	20.2	-	-	5.44	1.8	180
4. Liver residue meal	75.0	16.0	-	5.5	3.5					
5. Skim milk powder	40.0	0.2	-	54.0	8.8					
6. Yeast	48.0	1.2	3.0	40.8	7.0					
7. Meat meal	57.0	10.5	2.5	3.0	27.0	-	-	2.98	0.68	1.0
V. Roughages and vegetables wastes										
1. Lobia/cow pea	22.0	2.5	23.5	39.5	12.5	0.46	0.38			
2. Topioca	3.0	0.5	10.5	76.5	9.5	3.54	3.15			
3. Potato	9.0	0.5	2.0	85.0	3.5	3.74	3.00			
4. Carrot	10.0	1.5	9.5	69.0	10.0	-	-			
5. Turnip	14.0	2.0	12.0	62.0	9.5	-	-			
6. Mango seed	6.5	10.0	1.5	79.5	2.5	3.10	2.50			
7. Barseem	25.5	3.5	21.5	34.5	15.0	0.52	0.43			
8. Lucerne	20.5	2.5	23.0	44.0	10.0	0.53	0.43	0.7	0.31	0.8
9. Pea	12.5	1.0	30.0	48.0	8.5	3.25	2.60			
10. Guar	15.0	1.7	30.5	43.5	10.0	0.43	0.35			

Reference : Pathak and Ranjhan (1988) and Ranjhan (1980).

Nutrient Requirement of Hogs

Table 4.2 Nutrient requirement of growing swine fed Ad Libitum or amount per kg. diet (Based on NRC requirements - Anon, 1973 of U.S.A.)

	Weaning	*Growing*	*Finishing*
Live weight (kg.)	*5-12*	*12-50*	*50-100*
Daily gain (kg.)	*0.30*	*0.50*	*0.60*
Nutrients	*Requirements for different class of pigs*		
DE (M Cal/kg.)	3,500	3,500	3,300
M.E. (M Cal/kg.)	3,360	3,360	3,170
Crude protein (%)	22	18	14
Calcium %	0.80	0.65	0.50
Phosphorus %	0.60	0.50	0.40
Sodium %	-	0.10	-
Chlorine %	-	0.13	-
B-Carotene (mg)	4.4	3.5	2.6
Vit. A (IU)	2,200	1,750	1,300
Vit. D (IU)	220	200	125
Vit. E (mg)	11	11	11
Thiamin (mg)	1.3	1.1	1.1
Riboflavin (mg)	3.0	3.0	2.2
Niacin (mg)	22	18	10
Pantothenic acid (mg)	13	11	11
Vit B_6 (mg)	1.5	1.5	-
Choline (mg)	1,100	900	-
Vit. B_{12} (mg)	22	15	11
Amino Acids (per cent)			
Arginine	0.28	0.20	0.16
Histidine	0.25	0.18	0.15
Isoleucine	0.69	0.50	0.41
Leucine	0.83	0.60	0.48
Lysine	0.96	0.70	0.57
Methionine + cystine	0.69	0.50	0.41
Theronine	0.62	0.45	0.37
Tryptophan	0.18	0.13	0.11
Valine	0.69	0.50	0.41
Phenylalanine + Tyrosine	0.69	0.50	0.41

Table 4.3 Nutrient requirement of breeding Swine (Amount per kg. of diet or per cent)

Nutrients	*Breeding gilts and sows (wt. 110-160 kg.)*	*Lactating Gilts and Sows (wt. 140-200 kg.)*	*Boars (Young and adults (wt. 110-180 kg.)*
Crude protein %	14	15	14
D.E. M. Cal/kg.	3,300	3,300	3,300
Calcium%	0.75	0.6	0.75
Phosphorus %	0.50	0.4	0.50
NaCl %	0.50	0.5	0.5
ß Carotene (mg)	8.2	6.6	8.2
Vit. A (IU)	4,100	4,300	4,100
Vit. D (IU)	275	220	275
Thiamin (mg)	1.4	1.1	1.4
Riboflavin (mg)	4.1	3.3	4.1
Niacin (mg)	22.0	17.6	22.0
Pantothanic acid (mg)	16.5	13.2	16.5
Vit. B-12 (mg)	13.8	11.0	13.8

4.5 WATER REQUIREMENT OF PIGS

4.5.1. Provision of Water

Adequate, fresh, clean, odourless, pure and safe from parasitic ova or larvae.

4.5.2 Factors Affecting Water Intake

Ambient temperature, composition of feed, kind of feeds, preparation of ration/ feed, style of feeding, age, body, size, breed, season, health, cleanliness, physiological stage of animal.

4.5.3. Amount of Water Per Day/Hog

2 to 3 kg. per kg. of dry feed which in summer may be as high as 4 to 5 kg per kg of dry feed.

4.5.4. Supply of Water Requirement

(a) By voluntary intake of drinking water.

(b) Partly by succulant green fodders.

(c) Very little by oxidation of food stuffs in the tissues called metabolic water.

4.6 FEED MIXTURES FOR BREEDING STOCK

	Feed ingredients	*Mixtures*		
		1	2	3
1.	Maize	50	28	30
2.	Wheat bran	27	25	30
3.	Ground nut cake	15	20	20
4.	Damaged wheat	-	20	-
5.	Gram/Pulse chuni	-	-	8
6.	Fish meal	6	5	-
7.	Mineral mixture	1.5	1.5	1.5
8.	Common salt	0.5	0.5	0.5
9.	Antibiotics (gm)	20	20	20
10.	Rovimix (gm)	20	20	20

Amount to be fed : 3.5 to 4.5 kg per day, divided in three parts.

Note :

1. Louis and Lewis reported that boars fed a low level of protein take longer to mount a dummy and start ejaculating, and produce less sperms.
2. Boars with low protein have reduced libido-due to decreased oestrogen concentration in circulation.

4.7. RATION FOR FATTENING STOCK

Maize	25%
Gram	25%
Barley	23%
Linseed cake	25%
Mineral mixture	1.5%
Common salt	0.5%

Note : Composition of rations for different class of pigs may be seen in tables 4.3, 4.4 and 4.6.

4.8 PASTURE FOR HOGS

4.8.1 Advantages

1. Brings saving in grains consumption.
2. Helps is making feeding of hogs cheaper.
3. Increases profits from hogs.
4. Provides major protion of protein, minerals and vitamins to hogs and prevents nutrition deficiency due to vitamins and minerals.
5. Keeps hogs healthier.
6. Better growth of pigs.
7. Prevents anaemia in pigs.
8. Saves labour.

9. Conserves manure.
10. Increases rate of gain.
11. Favours limited feeding.
12. Improves reproduction.
13. Prevents erosion.
14. Needs less fertilizers.
15. No cost of fitting soil.

4.8.2 Suitable Pasture Grasses for Swine

Lucerne, red clover, ladino clover, sudan grass, cow pea, berseem.

Note : Small pasture is good for raising pigs provided it has succulant green forages and kept parasite free.

4.9 TYPE OF FEEDING : "SELF"

Good pasture can be used to replace concentrate mixture in swine as follows:

(a) 10 to 20 per cent of the concentrate for growing and fattening pigs.
(b) 20 to 25 per cent of the concentrate mixture for pregnant sows.
(c) 50 per cent of the concentrate mixture for pregnant sows.

Table 4.4. Recommended rations for different class of pigs.

Ingredients (per cent)	*Weaning pigs (5 - 15 kg) creep feed*	*Weaners/ Growers (15 - 50 kg) body wt.*	*Finishers ration (50 - 90 kg.) body wt.*	*Pregnant Sow/ Gilts*	*Nursing sow*	*Boar*
Maize	55	50	45	50	55	60
Ground nut cake	20	20	20	18	15	20
Wheat bran	10	18	25	20	18	13
Molasses	5	5	5	5	5	—
Fish meal/Meat meal	8.5	5	3	5	5	5
Mineral mixture	1	1.5	1.5	1.5	1.5	1.5
Common Salt	0.5	0.5	0.5	0.5	0.5	0.5
Antibiotics supple-ment (g. per quintal	30	20	15	-	10	-
Rovimix ($A+B_2+D_3$)	25 g	20g	15g	20g	20g	15g
Crude protein %	22	18	14	16	15	14
Nutritive ratio	1 : 4	1 : 5	1 : 6	1 : 5	1 : 5	1 : 5

When pasture is used to replace the concentrate mixture for swine, it is advisable to restrict the level of concentrate by hand feeding because pigs will not reduce the concentrate intake if these are self fed.

4.10 SUGGESTED REPLACEMENT FOR FEEDS

1. Maize by cereals like brown rice, wheat, bajra/ sorghum, barley.
2. G.N.C. by linseed cake, coconut, cake, seasame cake, soyabeen cake.
3. Fish meal by skim milk powder.

Note :

1. Vitamins are not necessary if pigs are fed fresh green legums like lecerne, cowpea, berseem.
2. Feed costs about 80% of pork production hence great care must be taken in computation of ration.
3. The ration must be economical, nutritious, capable of providing nutrients according of needs.
4. Incorporate vegetable wastes, left over feeds, refuse from hotels and restaurants, damaged food grains unsuitable for human consumption etc. of minimize cost of ration.
5. Fresh drinking water must be made available at all times.
6. Green leguminous fodder @ 1-2 kg per/day adult animal may be fed.
7. Pre starter pigs ration must have 24 per cent protein and low fibre content. Piglet start nibbling food when they are about 1 week old. They consume about 1.5 to 2 kg feed to attain 5 kg body weight.
8. Piglets are prone to anaemia therefore Sow's udder may be swabbed with iron sulphate mixed with sugar. Iron injection may be given on 4th and 14th day of age.

4.11 AMOUNT OF RATION FOR PIGS FEED/PIG/DAY

Class of Pig	*Live weight in kg*	*Feed to be fed in kg*
Growers	25	1
	26 to 45	2
	46 to 100	3
	Above 100	4
Pregnant Sow	150	3.5
Lactating Sow	150	5
Boars	150	3.5
Sows	150 to 225	4.5

4.12 MINERAL MIXTURE FOR PIGS

Mineral ingredients		*per cent*
Ground limestone	=	57.7
Bone meal (Steamed)	=	20.0
Iodized common salt	=	20.0
Iron sulphate	=	2.0
Manganese sulphate	=	0.2
Copper sulphate	=	0.1

4.13 FEEDING OF PIGLINGS

Age	*Weight gained*	*Daily feed in kg.*	*Cumulative feed in Kg*
Birth to 1 month	-	Mother milk	-
1 to 2 months	-	0.5	15
2 to 3 months	-	1.0	30
3 to 4 months	-	1.2	36
4 to 5 months	-	2.0	60
5 to 6 months	-	2.5	75
Total	70	-	216 kg.

4.14. FEEDING DIFFERENT CLASS OF PIGS

1. **Suckling pigs (up to 1 week age)** : These piglets are of 1 to 2 kg and nourised on sow's milk. They are helped to suckle about 8 to 10 times a day.

2. **Pre-starters (7 to 21 days age group) :** These pigs usually start nibbling at feed at 1 week age. Prestarter ration increases growth rate and reduces demand on sow's milk. Since milk production in udder of sow start declining after 3 week of lactation, it is necessary that piglings start on feeding solid food to meet nutritional requirement.

Pre-starter ration should have 24% protein and rich in vitamins but low in fibre. A proportion of antibiotics in ration of small pigs is recommended to protect from diseases. Piglings on pre-starter ration gains about 3 kg to weigh about 5.1 kg.

3. **Starter pigs (21 days to 35 days age group)** : Piglings between 3 to 5 weeks age grow fast. These are fed creep ration containing 18% protein. The total amount of creep feed required per pigling will be 9 to 11 kg to yield a gain of 6 kg so piglets weigh about 11 kg at 5 weeks of age.

This ration is self fed in a feeder separately from the mother sow. Feeders are placed inside creep where mother can not enter. Lot of fresh water is required for drinking by starter pigs.

4. **Pre-weaner pigs (age group 35 to 56 days)** : Pre-weaner pigs weigh

between 11 to 23 kg. Pigs during this stage consume 20 - 25 kg pre-weaner ration containing 16% protein to yield a net gain of about 12 kg between 5 to 8 weeks age. Pigs at 56 days of age are weaned and housed separately from sow mother.

5. **Growers and barrows (age group 2 to 5 months)** : During 2 to 5 months age the weaned pigs grow from 23 kg to 57 kg body weight, called feeder pigs. Access to pasture and supplementry feeding results in rapid growth. Growser ration must contain 14 to 16 per cent protein and self fed.

6. **Finisher pigs and stags (age group 5 to 8 months)** : These type of pigs are fed low protein (12%) but high energy ration for fattening to make them ready for marketing. The average weight of this class of pigs at 6, 7 and 8 months age should be about 60 - 70 kg, 70 - 90 kg and 120 kg, respectively.

7. **Feeding of boars** : Boars for breeding should be kept in a thrifty condition. Protein in boar ration should be about 16 per cent. Boar should not be too fat as at weakens leg and breeding ability. Boar should be given regular exercise to keep in good health.

8. **Feeding gilts** : Gilts should be fed with limited amount to avoid over fattening and keep them thrifty. If they are fat, there will not be any scope for flushing during breeding season. The ration given to gilt should be about 2/3 of the requirement of growing and fattening pigs, as limited feeding will increase the litter size and reduces feeding cost. Ration of gilt must contain 14 to 15 per cent protein.

9. **Feeding pregnant sows** : Ration must have 15 per cent protein and quantity will depend upon age, stage of pregnancy and condition of animal. Pregnant sows need about 2 kg feed per day. The sows in gestation should gain 27 to 36 kg compared to gilt 32 to 41 kg. About 25 to 30 per cent lucerne may be included in ration. Proper feeding of sow will result in -

(a) A larger litter (litter weight and number),
(b) Heavier and healtheir piglets at birth,
(c) Enough milk yield by sow,
(d) Reduction in still birth, abnormalities and stunted growth, and,
(e) More and heavier pigs weaned.

Note :

1. A pregant sow should not be overfed as over fatigue can cause problems.
2. Milk of over fat sows contain more fat and less protein.
3. Over fat sows dont's eat enough after farrowing which reduces milk production which affects the piglings.
4. Fat sows causes problems at farrowing time.

10. **Sows during and after parturition** : About 4 to 5 days before farrowing in sow's ration should be reduced to half. More laxative feed consisting of wheat bran must be given. On the day of farrowing no feed is given, but enough clean

water for drinking purpose be provided at all times. It is best to feed same bulky ration but only 50% of the quantity on first day of farrowing and then increase ration by 200 to 300 gm daily until she is fully fed.

11. Sows in lactations : Ration for sows of lactation will depend upon nutritional needs of piglings however 2.5 to 3.6 kg per day will be enough for small litter (8 to 10 piglets). A lactating sow requires 14 to 23 litres of water per day. The sow in full milk production requires 4 to 6 kg of feed per day depending upon weight, its condition, milking capacity and size of litter.

Note :

1. Weaning weight and vigour of the piglings depend on care and feeding of lactating sow.
2. Gain in weight of suckling piglets is directly related to the milk producing ability of the mother.

12. Unbred or dry Sow : Unbred sow requires 14% protein in their ration. If they are lean and run down in condition due to previous pregnancy and lactation it requires more nutrition otherwise such sow requires ration only for maintenance.

Summarized guide line for feeding market hogs of different groups

Class of pigs	*Protein % in ration*	*Weight group*	*Age group*	*Average body weight*	*Total gain in weight*	*Remark/ Feeder Space*
Suckling	mother's milk	1 - 2 kg.	0 - 7 days	2 kg.	1/2 - 1 kg	Sucking 8 - 10 times a day.
Prestarter	24%	2 - 5	7 - 21 days	5 kg	3 kg	—
Starter	18%	5 - 11	21 - 35 days	11 kg	6 kg	10 cm space feeder 15 cm. space
Pre-weaner	16	11 - 23	35 - 56 days	23 kg	12 kg	
Growers	14	23 - 57	57 - 90 days	35 - 40	15 kg	
			90 - 120 days	40 - 45	10 kg	
			120 - 150 days	50 - 60	10 kg	
Finishers	12	60 - 70	5 - 6 months	70	12 kg	
		70 - 90	7 months	90	20 kg	
		90 - 120	8 months	120	30 kg	

4.15. EFFECT OF DIFFERENT TECHNIQUES OF PREPARATION OF FEEDS

1. Cooking : It improves the food value of feeds like soyabeans, potato, beans etc., but in general feed value of most feeds decreases rather than increase by cooking.

2. Fermenting : Fermenting feeds with culture like yeast does not cause any significant increase in feeding value. Moreover amount of labour required is more.

Note : If ration is deficient in vitamins yeast fermentation of feed increases vitamin B complex value in it.

3. Grinding/milling : It increases the digestibility by 20 per cent but not the nutritive value. Therefore medium to coarse grinding of small grain such as oat, barley, rye etc. is usually profitable. Although extent of saving from milling will depend upon method of feeding (hand vs self fed) and hardness of seeds. Following should be the fineness of grinding of swine feeds :

Swine feed	*Extent of fineness desired*
Soyabean	Very fine
Barley grain	Medium
Oats	Medium
Rye	Medium
Wheat	Medium
Maize/corn	Shelled.

Note : All other grains must be coursely ground as very fineness reduces palatability, increases possibility of caking. Too fine ground grain is not suited to self feeding.

Soaking/wet feeding

Soaking is not needed because it improves neither palatability nor digestability.

Merits of wet feeding

(a) Increases water consumption.
(b) Prevent wastage of feed.

Demerits of wet feeding

(a) Inconvenience in mixing of feeds.
(b) Extra labour and time.
(c) Sloppy feed does not run well in self feeders.

Note :

1. Wheat bran may be mixed with warm water and fed as grain mash.
2. For soaking grain mixture use of skim milk of butter milk is more advantageous.

4.16. FEEDING ANTIBIOTICS TO PIGS

Feeding antibiotics as feed additives is important for following reasons :

1. Helps in success of intensive system of swine production.
2. Reduces incidences of scouring and unthriftiness in pigs.
3. Causes 10 to 15 per cent more rapid gain and feed conversion.
4. Enhances growth rate in early age of piglets.
5. Increase feed efficiency of growing and finishing pigs up to 5 per cent.
6. Reduces the number of runts and thereby make more uniform crop of pigs.

Types of antibiotics for pigs

Penicillin, bacitracin, auromycin (Chlorotetracycline), terramycin. etc.

Amount

10 to 20 mg per kg. feed depending upon types of antibiotics.

Choice of antibiotics

It will depend upon cost of antibiotics relative to that can be expected, the general disease level of the herd and other factors.

4.17. FLUSHING OF SOWS/GILTS

4.18. FEEDING SWINE FOR MARKET

Preparation of hogs for market starts at an early age consisting of two stage *viz.,* growing and fattening. There are two methods of feeding of finishing hogs :

(a) Full feeding all the time until animal reaches its market weight.

(b) Limited feeding in growing period followed by full feeding in the last 2 to 2.5 months period before marketing.

4.19. THE COMPOSITION OF RATION

(a) Cereal based ration :

Maize	=	25 per cent
Sorghum/Barley	=	22.5 per cent
Rice polish/wheat bran	=	15 per cent
Ground nut cake	=	12 per cent
Til cake	=	10 per cent
Fish meal	=	10 per cent
Min mix	=	5.5 per cent
Rovimix ($A + B_2 + D_3$)	=	10 g

(b) Non-cereal ration :

Wheat bran	=	70 per cent
Ground nut cake	=	12 per cent
Til cake		

Fish meal	=	5.5 per cent
Mineral mixture	=	2.5 per cent
Rovimix	=	10 g
$(A + B_2 + D_3)$		

Note :

1. Leguminous pasture may also be advantageously used in both methods of feeding (limited and full feeding), depending upon following factors :

   (a) Prices of feeds.
   (b) Feed availability.
   (c) Market condition.
   (d) Quality of pasture.
   (e) Extent of pasturage available.
   (f) Labour.

2. Self feeders are quite suitable in full feeding.
3. On an average with feed efficiency ration 1 : 3, the weight gain of 70 kg. in six months period be attained by consumption of 216 kg. feed.
4. Plane of nutrition in latter phase of finishing period affect the fat production in hogs. Therefore to produce leaner hogs (for lean meat production), the restriction of feed intake is done when pigs are of 4 months age or of 50 kg. body weight. Restriction in feed intake causes decrease in weight gains. Every 1 per cent restriction in feed causes 0.53 per cent decrease in black fat thickness.

5

Management Practices for Hogs

5.1 BODY PARTS OF PIG

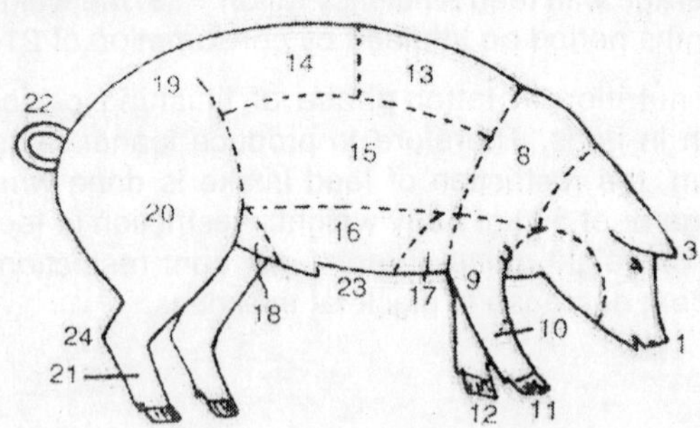

Fig. 5.1 Body parts of hog.

The above figure shows the parts of a hog : 1. snout; 2 eye; 3 ear; 4. face; 5. jowl; Є. forehead; 7. neck; 8. shoulder; 9. foreleg; 10. pastern; 11. toe; 12. dew claw; 13. back; 14. loin; 15. side; 16. belly; 17. foreflank; 18. hind flank; 19. rump; 20. ham; 21. hind leg; 22. tail; 23. sheath; 25. hock.

It is important to know body parts as it is useful for following :

1. Study of breed characteristics.
2. Judging animals.
3. Locating external abnormalities on body.
4. Writing report of treatments and dressing of parts etc.
5. Proper classification of market hogs.

5.2 SWINE JUDGING

A meat hog should be deep, straight in the underline, reasonable long, wide and even in width, slightly arched and strong in the back, have straight legs of medium length, deep and full in ham and with a high tail setting. The chest should be wide deep, hair coat must be fine, straight and free from wrinkles. A hog should be just fat enough to be firm. Sows should have well developed udders with 12 or more teats. The ham should be broad and full. The ribs should be well sprung. A breeding hog should have strong legs and short pastern. He should be able to walk freely and gracefully without stiffness.

5.3 HANDLING PIGS

Operation	*Tool/technique used*
1. Handling	Pig catcher
2. Handling	Running noose slipped over upper jaw behind tusks.
3. Controlling	Nose ring
4. Casting	

Fig. 5.2 Pig Catcher

(a) Materials

(i) Two short ropes of 1 metre length each.
(ii) Two thick ropes of 5 metres length each.

(b) Procedure

(i) Tie both hind legs with a small rope above the fetllocks together.
(ii) Tie both forelegs with another small rope above the fetlocks together.
(iii) Fasten a 5 m. long rope's one end to the rope between forelegs and pass the other end of this rope through hind legs.
(iv) Fasten one end of the other 5 metres long rope between hind legs and pass the other end of this rope between forelegs.
(v) Two long ropes are pulled by two persons opposite to each other throwing **animals on ground.**

5.4 AGE DETERMINATION

Dental Formula of Teeth in Pig

Jaw	*Incicors*	*Canine*	*Premolars*	*Molars*	
Upper	3 + 3	1 + 1	4 + 4	3 + 3	= 44
Lower	3 + 3	1 + 1	4 + 4	3 + 3	

Temporary

Eruption of all temporary teeth is complete by 5 months age.

Permanent

First pair (Corner)	by	6 months age.
2nd pair lateral	by	10 months age.
3rd pair middle	by	12 months age.
4th pair central	at	1.5 years age.

5.5 CASTRATION IN PIGS

Purpose

1. To eliminate undesirable males.
2. To prevent indiscriminate breeding.
3. To make animals more docile.
4. To prevent boar odour in cooked meat.
5. To develop pork to superior quality
6. Important in treatment of accidental injuries tumour of testes, etc.

Kind of pigs : males not selected for breeding.
Age of piglets : 3 to 4 weeks age.
Method : Operation method (Fig. 5.3)

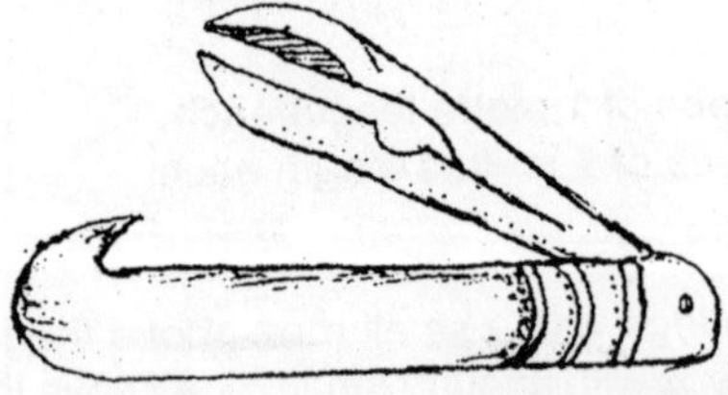

Fig.5.3 : Castrating knife with double blade (scalpel and hock).

1. Secure the piglet and make it lie down on dorsal side (up side down) and grasp feet under control.
2. Make hands and castrating knife clean and sterilize with spirit soaked cotton.
3. Wash the scrotum with an antiseptic solution of acriflavin.
4. Make an incision on the scrotum on each side deep enough extending the cuts well down to permit proper drainage.
5. Remove the testes with its membrane by pulling them backward, and bring with them as much cord possible and break it by cutting.
6. Scrotum and surrounding parts must again be disinfected thoroughly with tincture iodine.
7. Apply little sulphanelamide powder mixed with iodoform (fly repellent) to prevent infection or other complication.

5.6 IDENTIFICATION

Purpose

1. To maintain proper records of pigs.
2. To ensure proper care in feeding of pigs.
3. To have better management practices wherever needed.
4. To designate animals properly.

Procedure

Ear notching is the common method used for marking and identification of pigs, which is described as follows (Fig. 5.4) :

1. Procure and hold the piglets.
2. Sterilize the side ear punch and central ear punch or pair of sharp scissors or pincers.
3. Clean the ear with the help of cotton spirit.
4. Side ear nothces be made by punch.
5. In case of hole if needed make use of sterilized central ear punch.

Note : Care should be taken not to make nothces too small to close up soon and not too large to.deform the shape of ear.

Fig. 5.4 Notch methodof marking in pigs

5.7 PIG EXCERCISING

Purpose

1. To keep animals fit and growing
2. To maintain good appetite
3. To prevent stiffness in the limbs of pigs.
4. Desirable to maintain good digestion.
5. To prevent much difficulty at farrowing.
6. To help sow to produce more vigorous piglets.
7. For general well being and good health.
8. To make them docile.

Amount of exercise : Limited but ample each day.

Method

1. By providing free run–out in uncovered area.
2. By providing free access to pasture.
3. Millen (1947) suggested that the exercise pen with a small opening on central platform is 2.5 m wide and 7.5 m long. Each pen is quite open at the top but surrounded by wall of bricks upto 1 m in height.

Note :

1. Avoid strenuous exercise specially on sunny day as it may prove to be fatal.
2. Handle all animals gently and kindly.
3. Handle tempramental/nervous animals with care.
4. Never keep two boars together.

5.8 PREPARING CALENDAR OF DAILY OPERATION

Pupose

1. To get pigsty jobs completed on time and properly.
2. To utilise labour efficiently.
3. To give better care to animals.
4. To get higher returns.
5. To get jobs done in orderly manner.

Timings	*Routine and periodical farm operations*
7:30 to 10 A.M.	**I. Daily jobs** 1. Put breeders stock on pasture. 2. Cleaning all pig houses and disinfection. 3. Cleaning of farm premises. 4. Segregation of sick animals. 5. Isolation of sows and gilts "in–heat" 6. Feeding half of the daily feed (breeders limited feeding). 7. Feeding free choice ration to all growers and finishers in self feeders (full feeding system).
	II. Other periodic jobs 1. Weighing of stock. 2. Treatment of sick animals. 3. Removal of needle teeth in baby pigs. 4. Marking of piglets for identification. 5. Vaccination of pigs.
10 to 12:00	**I. Daily jobs** 1. Bringing breeding stock from pasture back to sties. 2. Removal of manure and conservation. 3. Wallowing hogs or washing pigs with hose pipe.

Timings	Routine and periodical farm operations
	II. Other periodical jobs 1 Preparation of concentrate mixtures. 2. Spraying insecticides. 3. Treatment for mange/scabies etc. 4. Castration of animals. 5. Sale of animals.
12.00 to 13.30	**Lunch Time**
13.30 to 17.00	**I. Daily jobs** 1. Supply remaining half of daily feed to breeders (limited feeding). 2. Marking entries in farm records. 3. Shifting nursing sows in farrowing pens. 4. Cleaning the houses. 5. Cleaning and disinfection of farrowing pens. 6. Wallowing hogs. 7. Transfer gilts and sows approaching parturition to farrowing stalls. **II. Other periodic jobs.** 1. Taking care of farm purchases. 2. Deworming of hogs. 3. Cleaning, scrubbing and whitewashing of feeding and water troughs weekly. 4. Weaning of piglets. 5. Making provision of rotation of pasture. 6. Assignment of jobs for next day.

5.9. HOG WALLOWS

Pig need a wallow during summer because they have relatively few sweat glands.

It is water pool where hogs enter into water and cool their body in summer. It is specially needed for fattening and breeding animals.

Location : In a shady place.

Proper location of wallow in relation to sites is important so that during treatment the hogs should not have access to any shallow pools or muddy holes which they usually prefer.

Type of wallows

A masonry wallow made of concrete and cement with proper drainage system be made. Mud wallow is not at all desirable as it causes insanitary conditions.

Size of wallow

It depends upon size and number of animals. It is advantageous to have large wallow enough to accomodate entire herd at one time. A wallow of 3 metres length, 2 metres width and 45 cm. depth will easily hold a herd of 20 hogs of various ages. Where sprinklers are used one nozzle per 25 to 30 hogs is sufficient.

Construction

The top of the side and end walls of wallows should extend at least 10 cm above the surface of ground to prevent surface water running into wallow. The entrance and exit incline must have a gradual and easy slope so that hogs may enter and leave wallow conveniently. To prevent slipping, the concrete floor must be roughend with a broom soon after it is laid and grooves made in it with a bar.

Medicated wallows

It is used for medicating or disinfection purposes. The instinctive habit of the hog to wallow in water when weather is warm, can be advantageously used in treatment for external parasites and diseases caused by them (mange, scabies, hog louse). Proper depth of liquid in wallow depends upon size and number of hogs. A 8 to 10 cm liquid will give better results than greater depth. Wallow should not be kept medicated continuously as disinfectant may cause irritation with the result hogs may refuse to enter the wallow. It should be drained, cleaned and recharged with water only. Disinfestant may be added every week or 10 days until desired results are obtained.

Precautions

Hogs are likely to drink from wallows unless water is rendered denatured by distasteful substance. For this crude oil and petroleum products are most suitable dips because of distasteful nature and tendency to spread over surface of the body. Crude oil as disinfectant may be used @ 1 pint per each hog. The oil floating on surface also prevent evaporation of water.

5.10. HOUSING OF HOGS.

5.10.1 Objectives

1. To provide, shelter at a cheaper cost.
2. To protect animals from theft.
3. To protect pigs from sun heat, hot and cold winds and other adverse conditions of inclement weather.
4. To provide clean and comfortable shelter.'

5.10.2. Advantages of Adequate and Proper Pig Sties

1. Saving and better use of labour.
2. Increased meat production.
3. Better health of animals.
4. Decrease in mortality of piglets.

5. Control of parasites.
6. Prevents of diseases.
7. Better care and supervision of hogs.
8. Better reproductive efficiency in pigs.
9. Proper controlled feeding of animals.
10. More returns
11. Encouragement to other hog raisers.

5.10.3. Factors Affecting Construction and Location of Pig Houses.

1. Space available.
2. Materials available.
3. Capital/financial position of farmer.
4. Soil–fertile for pasture and sandy loam.
5. Elevation–higher than surrounding for effective drainage.
6. Topography - somewhat levelled.
7. Sunlight–for exposure of floor to maintain clean, but shady place for animals.
8. Protection for wind–to avoid direct draft of heat and cold wind on animals.
9. Accessibilty – Easily accesible to field and roads
10. Market–close by for easy disposal of pork and procure goods.
11. Labour–Suitable for this class of animals
12. Surrounding–Avoiding such communities that have religious taboo for this class of animals.
13. Electricity–be available enough and continuous for forced ventilation.
14. Miscellaneous – Efficiency of labour, fire protection, appearance, etc.

5.10.4. Desired Basic Requisites of Pig sties

1. To give maximum comfort for better growth.
2. To avoid dampness in houses.
3. To protect from draft and overheating of sties.

5.10.5. Flooring

1. It should be made of cement and concrete with grooves to prevent slipping; impervious, easy to clean with rough surface.
2. Provide enough slope from one end to the other for easy drainage (about 3 cm for every 2 metre length)
3. Provide enough shade by planting trees.

Floor space requirement for different class of pigs

Category of pigs	Roofed/covered area (m^2)	Open yard area (m^2)
1. Boar	6.0 to 7.5	8.8 to 12.0
2. Farrowing sow	7.5 to 9.0	8.8 to 12.0
3. Weaner/fattening pigs	0.9 to 1.8	.9 to 1.2
4. Dry sow/gilt	1.8 to 2.7	1.4 to 1.8

Floor should be made for conventional masonry type with cement mortar. Proper drains be provided for disposal of effluents.

5.10.6. Waterers and Feeders Requirements

Particulars	Adult pigs	Growing pigs
1. Length of manger/pig (cm)	60–75	25–35
2. Length of water trough/pig (cm)	60–80	30–40
3. Width of manger and water trough (cm)	50	30
4. Depth of manger and water trough (cm)	20	15
5. Height of manger and water trough (cm)	25	20

Waterers and feeders should be kept inside the shed. The amount of water for boar, farrowing sow, fattening pigs and dry sow will be about 40 to 50, 18 to 20, 3.5 to 4 and 4 to 5 litres, respectively. Water needed by the hogs will depend upon following factors :

1. Age and body size.
2. Ambient temperature.
3. Breed.
4. Adaptation.
5. Season/weather.
6. Humidity.
7. Dry matter in feed.
8. Number of suckling pigs.
9. Frequency of water supply.
10. Quality of feed.

Kind of water needed : Clean, odourless, pure, colourless, tasteless, free from toxic substances, free from germs and parasitic ova/larvae.

Walls of pig houses

(a)	Solid brick wall	=	60 cm
(b)	Railings of tubular iron (pipe) above brick wall	=	140–190 cm
(c)	Total height of walls for roofing	=	200–250 cm

5.10.7. Roofing of Sties

Roofing of sties can be of either asbestos, zinc sheet, tiles or coconut leaves, etc. Height of the side walls must be 2 to 2.5 meters.

Doors :

Doors 2 x 1 m size be made strong and fitted close to floor so that pigs may not lift it up by putting snout under it.

5.10.8. Guard Rails/ Pig Fenders

Purpose : To prevent loss of baby pigs from crushing by sow in farrowing pen.

Material : Galvanized iron pipes of 5 cm diameter.

Construction : Height of pipes above ground should be 25 cm, and 20 cm from the side wall of farrowing pen. Distance ebtween pipes should be 10 cm.

Note :

1. It is usually the less active pig that is crushed.
2. Piglings can run under these rails when sow lies down so that they are not cruhsed between sow and wall.
3. There is less need for rails if pigs brooders are used.

5.10.9 House for Pigs

1. **Boar sty** : Keep these sties dry, vetilated with enough sleeping area. Only one boar be kept in one pen. Arrange all boar sties in row under one roof.
2. **Dry sow/gilt sty** : Keep 3 to 10 pigs in one such sty. Dry sows and gilts can be housed together.
3. **Farrowing pen** : Pregnant sow/gilt is kept here two weeks before farrowing. It should be large enough to accomodate sow and its piglings to move about up to nursing period (12 to 14 days).

Size 7.5 to 9m^2 or 8' x 10'

Note :

1. Desirable temperature 20°C.
2. Electric light bulb are fixed when heat is needed in winter.

Slope : When floor of such pen has slope the sow lies down with her back towards high side of pen and baby pigs will move towards lower side. This also reduces crushing losses.

Bedding : 10 to 15 cm bedding is needed specially in winter for good results. It must be kept clean dry and evenly distributed.

4. **Fattening pig sties** : About 20 to 30 pigs can be kept in one pen. These are meant for pigs raised for market.
5. **Weaner sty** : It is meant for piglets after 8 weeks of age upto 6 months. Not more than 30 pigs be kept in one such sty.

5.10.10. SYSTEMS OF PIG HOUSING

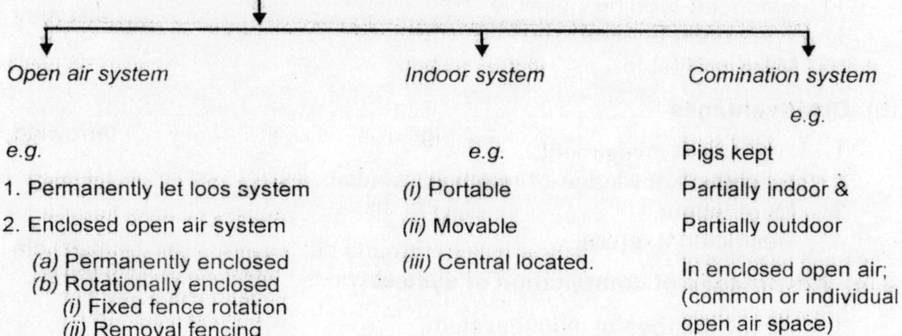

1. (a) Advantages of open air system

1. Cheapest method of rearing.
2. Low investment, only for make shift night shelter.
3. Less labour as no daily cleaning is needed.
4. Availability of plenty of sunshine.
5. Freedome of movement.
6. Benefit of exercise.

(b) Disadvantages

1. Pigs get no balanced feed.
2. Controlled breeding not possible.
3. Insanitary conditions.
4. More chances of diseases.
5. High mortality rate.
6. Exposure to adverse weather.
7. Reduced growth rate.
8. Lack of care to pregnant and lactating sows.
9. More area requirement.
10. Prone to attack of parasites.
11. Low profit.
12. Narrow bone to meat ratio.

2. (a) Advantages of indoor system

1. Better care as per requirement of pigs.
2. Proper controlled breeding possible.
3. Better sanitation.
4. Diseases occurrence minimized.
5. Less mortality.
6. Better growth rate.
7. Better care and supervision.
8. Parasitic infestation minimised.
9. Less area needed.
10. Controlled weather conditions.
11. Balanced feeding possible.
12. Wide bone to meat ratio.
13. More profitable.

(b) Disadvantages

1. High initial investment.
2. Scientific knowledge of rearing essential.
3. More labour.
4. Restricted exercise.

3 (a) Advantages of combination of system

1. All advantages of indoor system.

2. Cheaper than indoor system.
3. Enough sunlight and fresh air.
4. More area for exercise.

(b) Disadvantages of Combination system

All demerits of indoor system :

1. **Resting area** : Size per adult pig is 1.5 x 0.6 m however actual size should be consonance with feeding and exercising place. Pigs prefer to rest here most of the time.

2. **Feeding and watering space** : Size of feeding trough is 30 x 30 x 15 cm for length, width and depth, respectively. The size of the waterer is same as for the feeder, and can be constructed together in a line with 10 to 15 cm thick partition. Feeders can be constructed in one row.

3. **Dunging and exercising area** : It should be same as the feeding area but double the space of resting area in combination system.

4. **Drainage** : There should be proper drainage facilities from resting, feeding and dunging area to a common drain.

5. **Waste disposal area** :Two rectangular pits of 3 X 2.5 X 2 m will be sufficient for about 50–60 adult pigs for disposal of dung and other wastes. When one is filled, start filling the other. The decomposition takes place within 5 to 6 months, hence the size of pit should be such that it should take about 6 months to fill.

6. **Feed store** : A full grown pig requires 2 to 3 kg feed per day. Based on this requirement of feed and store can be estimated. If feed is to be purchased and stored for a month, the storage space should be sufficient.

7. **Recording facilities** : Various records on pedigree, breeding, feeding production, purchase and sale, ledger and accounts in the record room be maintained.

8. **Equipment and tools store** : The equipments and tools for cleaning, watering, feed preparation, feeders, waterers, disinfection, castration and slaughtering etc. are maintained in proper condition for use. The size of it will vary with the size of piggery.

9. **Medicine and first aid room** : A small shelf with essential tools and items will be enough for small piggery, whereas a hospital with proper staff may be needed in a big commercial piggery.

10. **Slaughter house and weighing space** : Weighing of pigs is done periodically to record growth rate. There is weigh bridge need to be arranged or economically feasible device to weigh feed and small piglets be made readily available.

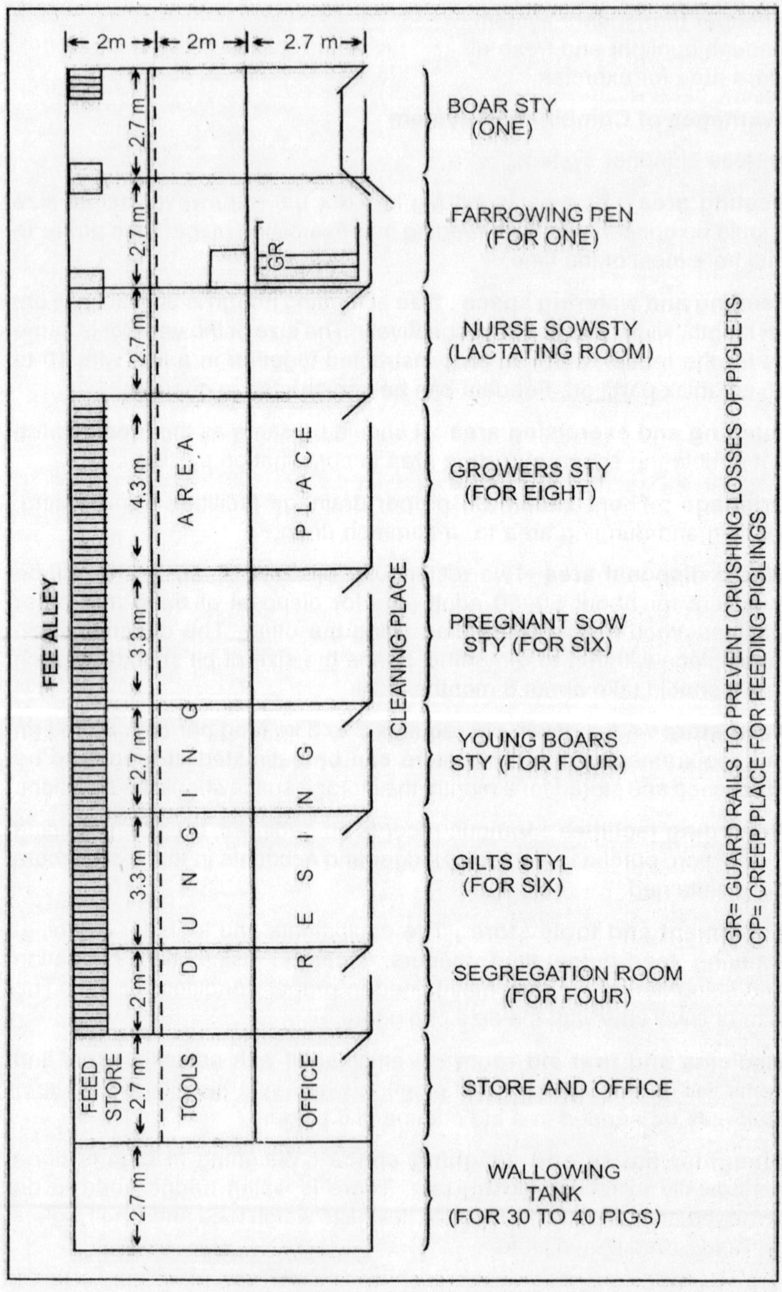

Fig. 5.5 Model Piggery Plan

Slaughter house consisting of balance, overhead rails for hoisting pig, tank for scalding water, platform for cleaning, dressing and preparing cuts from pork carcass is needed.

11. **Office facilities** : Piggery office facility can be combined with store/ record room/tools and medicine room but for a commercial farm proper office facilities are required for carrying out various transaction and managerial purpose.
12. **Feeding passage** : The feeders can be constructed on one side of feeding space. A feeding passage of 1.3 m in width can be constructed in such a way that feed can be distributed in feeders on one line.
13. **Cleaning passage** :Similar to feeding passage, a cleaning passage on opposite side of each room will be constructed close to outer wall.

5.10.11. Types of Houses

(i) Portable *(ii)* Movable *(iii)* Centrally located.

Shape of Houses

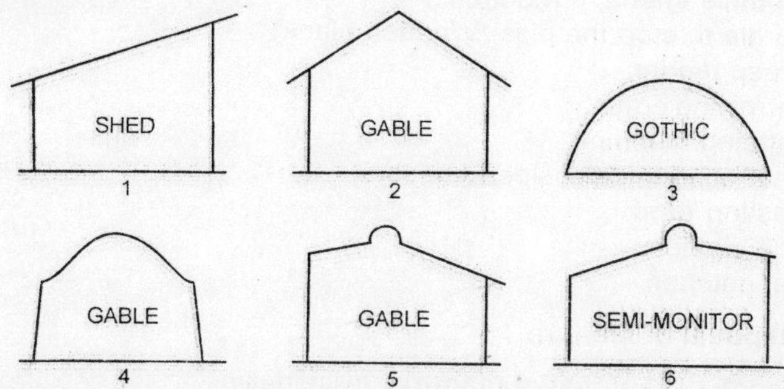

5.10.12. Advantages of Centrally Located (fixed) Houses Under one roof.

1. Better sanitation due to smooth floor and walls.
2. Saving in time and labour requirement.
3. More durability.
4. Better planned ventilation possible.
5. Hearing mechanism in cold weather could be installed.
6. Close attention to pigs.
7. Better care and supervision by direct operations from a central point.
8. Less need of equipments.
9. Better advertising value.
10. Securing better lighting arrangement.

5.10.13 Equipments for Swine Husbandry

1. Self feeders (Feeding troughs)
2. Feed guards.

3. Automatic trough covers.
4. Hog waterers
   (a) A simple barrel hot waterer.
   (b) Commercial barrel hog waterer.
   (c) Pressure system long waterer.
5.. Hog wallows.
6. Movable hog wallow.
7. Medicated hog wallow.
8. Dipping vats.
9. Electric pig brooders.
10. Breeding crate.
11. Combination crate for ringing, leading, vaccination, ear notching/tagging, etc.
12. Hog crate.
13. Fences.
14. Gaurd rails.
15. Pig catcher.
16. Protable sheds, if required.
17. Hurdle to stop the pigs (Wooden plank)
18. Creep feeder.
19. Farrowing crate.
20. Tattoing machine.
21. Castrating knife/scalpel, scissors.
22. Casting rope.
23. Bolt clipper
24. Ear notcher.

5.10.14. Disposal of Manure

Frequency of removal of manure : twice daily.
Methods are as follows :

(a) Solid manure : by means of wheel barrow and shovel, disposel into a pit for decomposition. Such manure will return 75 per cent of its fertilizing value to soil. Manure pits should be about 200 metres away in a place where no foul smell would pass through buildings. The production of manure from each adult hog is about 2 kg per day. The volumetric capacity of fresh manure is 700 to 900 kg/cu.m.

(b) In a liquid form by means of lagoon : Lagoon is a body of water like a small pond where in liquid manure is discharged and digested by bacterial action. In this method fertility value of manure is wasted but helps in saving of equipment and labour which may compensate the loss.

Pens are scraped and washed daily with water under pressure 75 lbs./sq inch and 500 gallons water per hour. This is run into lagoon which should hold at least one week accumulation of manure @ 2 kg/hog/day.

Bacterial action in lagoon

(i) **Aerobic** : By aerobic bacteria in the presence of air/oxygen.

(ii) **Anaerobic** : By green algae which use carbon dioxide, nitrate and other nutrients and in turn gives off oxygen for acrobic bacteria for oxidising waste materials. Anaerobic bacteria also take over to decompose the waste materials which may produce undesirable odour too.

Water in lagoon : Should be kept filled with water
Depth of lagoon : Approximately two metres.
Size of lagoon : @ 50 sq.ft./adult hog.

Location : 200 metres away from sites and prevent direction of prevailing winds.

Precaustions

1. Provide safety fence around lagoon to make dog and child proof.
2. Make the bottom levelled and impervious.
3. Clean the lagoon once in 5 to 8 years or when necessary to remove accumulated sludge if filled upto depth of one metre.

5.10.15. Disinfection of Pig Houses

Adopt following procedure :

1. Scraping floor and walls and remove manure completely.
2. Sweep thoroughly manager and feed troughs and scrub with brush suing hot water lye solution.
3. Burn all scrapings and sweepings.
4. Scrub drinking water trough with disinfectant solution.
5. Whitewash all walls, partitions etc., using 1/2 kg lime in litres of water.
6. Desinfect feed racks and gutters.
7. Dispose off all bedding and manure.
8. In case of earthen floor remove 5 cm top soil and replace it by new soil.
9. Provide one separate stall and runway for diseased hogs.
10. Expose floor to sunlight.

5.11 Sanitation and Disinfection on Hog Farms

Sanitation

It is the process of adopting hygeinic measures which nullifies the influences that deteriorate health and create conditions that secure health and ensures production of good quality products.

Problems due to inadequate Sanitation

Various problems is practical implementation for prevention of diseases are due to the fact that the swine rearing is managed by illiterate and ignorant persons who ignore the basic principles of hygenie and sanitation. Majority of causes for spread of diseases could be ascribed to improper sanitation which gives shelter to carriers of germs.

Importance

Proper cleaning and sanitation removes most of germs and parasites alongwith dirt, thereby remaining germs are few in number and possibly in weakened condition so as to be harmless under ordinary conditions. Following are the main points in the context :

1. Proper sanitation discovers causes of all preventable diseases.
2. It helps to devise means of rendering the cause ineffective if not removal of the causes of spread of diseases.
3. Helps in providing the most favourable conditions of life in respect of water, air, well sanitized sites, etc.
4. Helps in increasing the efficiency of animals.
5. Prevents economics losses due to infection.
6. Helps in development and growth of animals, making life vigorous and productive.
7. Lowers the rate of mortality and increases the longevity of animals.
8. Prevents occurrence of diseases and establishes conditions that ensure preservation of health.
9. Helps in minimizing contamination and production of good quality pork and pork products.
10. Helps hog man to learn and make continuous efforts for improvement.

Sanitizing Agent

It is a solution which will hold the number of bacteria below 4 bacteria/cm2 on the surface of utensils. In general, half the strength of a disinfectant substance is needed of sanitizing purposes.

Example : Disinfectant consisting of quaternary ammonium compounds may be diluted with only 60 per cent water for sanitizing purposes.

Sanitation Programme

It includes the following :

1. Adequate ventilation.
2. Proper lighting.
3. Adequate drainage.
4. Proper cleaning.
5. Proper disinfection.

Losses among hogs from infectious diseases and parasites often can be prevented if the following essential features of adequate sanitation are adopted in the living quarters :

1. Proper ventilation without drafts and without accumulation of moisture on walls and ceiling.
2. Proper disposal of manure, feed wastes and other excreta twice daily and keeping manure pit covered with straw to prevent breeding place of flies.

3. Proper construction of smooth and wide enough gutter for holding accumulated droppings and with proper slope to facilitate drainage to liquid excreta.
4. Watering and feeding utensils so constructed that they may be easily cleaned and thoroughly disinfected.
5. Good lighting programme through doors, windows, ventilators and artificial lights to facilitate proper cleaning and keeping floor dry.
6. Inside of walls smooth with corners rounded to facilitate cleaning and disinfection.
7. Use of proper and clean bedding material (*e.g.* saw–dust, wheat bhusa, paddy straw etc.) and removed at least once daily.
8. Avoiding use of permanent pastures where internal parasites or their intermediate hosts are found.
9. Judicious use of such insecticides that have no adverse physiological effects on the animal body, *e.g.*, malathion dust.
10. Adequate cleaning prior to effective disinfection.
11. Sweeping and scrubbing all feed troughs and passages and disinfecting with lye solution.
12. Burning of all sweeping and scrapings.
13. Application of heavy coating of white–wash containing a reliable disinfectant to the floors, walls and partitions, manager *etc.* (1/2 kg of lime in once gallon of water + disinfectant).
14. In case of mud floor a top of 12 to 15 cm soil is removed and replaced with clean soil.
15. Providing plenty of shade in hot weather.
16. Routine programme of deworming specially for animals on pasture.
17. Judicious spraying for lice and mange.
18. Segregating the sick animals.
19. Protecting feed and water from being contaminated with sewage disposals.
20. Proper disposal of infected litter and carcass.
21. Proper cleaning and disinfection of farrowing pens.
22. Abundant supply of clean water with good pressure for easy and effective cleaning
23. Cleaning should be followed by the use of disinfectant over all surfaces.

DISINFECTION

Disinfectant

Compounds used to kill bacteria and parasites are called disinfectant. The terms disinfectant, germicides-bactericidal agents, are based upon the effects of the commonly used concentration on the vegetative cells of pathogenic bacteria.

Disinfestant

Some parasiticides or insecticides which destroy animal parasites such as lice, mites, ticks and fleas are termed as disinfestants, used externally under sanitary control programme. Many but not disinfectants are also destructive to these external parasites such as coal–tar, petroleum, nicotine sulphate solution, sodium fluoride.

Disinfection

It means act of destroying the cause of an infection. Since the causative agents of many diseases are extreme small and may remain indefinitely in dust, cracks and crevices of buildings disinfection must be carried out carefully to eradicate common enemies of life such as bacterial, viruses, molds, and eggs of insects from contaminated premises.

Methods of Action of disinfectants

1. Destruction of bacterial cell or disruption of its organisation.
2. Interference with energy utilization.
3. Interference with synthesis and growth.

Types of Disinfectants

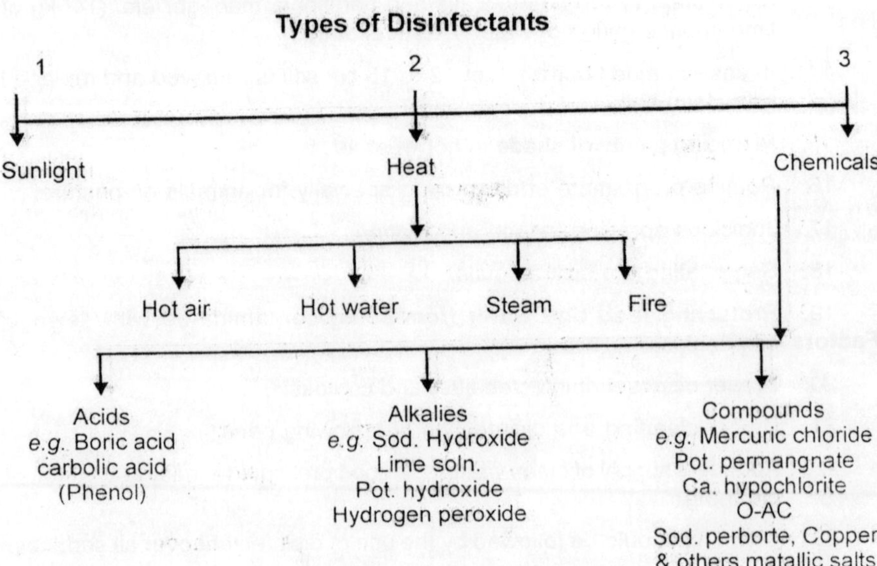

1. Sunlight

It is often a valuable disinfectant if surface are exposed directly for a sufficient duration. It loses its power to kill germs after it passes through thin film of water, dust or ordinary glass. Nevertheless well lighted houses for animals is of great importance. The disinfecting action of it is due to ultraviolet rays.

2. Heat

1. **Hot Air** : It is an effective means of disinfection but often an expensive once, hence is limited to laboratories.
2. **Hot Water** : Almost all utensils can be disinfected by immersion in boling water for a little more than 5 minutes. It is not satisfactory way for disinfecting floors as it loses its heat soon.
3. **Steam** : It is a satisfactory mean of disinfection but being expensive its use as disinfectant is chiefly limited to utensils. It is used under 15 lbs pressure.
4. **Fire** : It adds to the total destruction of bacteria and spores, therefore a best means of disposing infecting carcass and litter.

3. Chemicals as Disinfectants Kill germs by

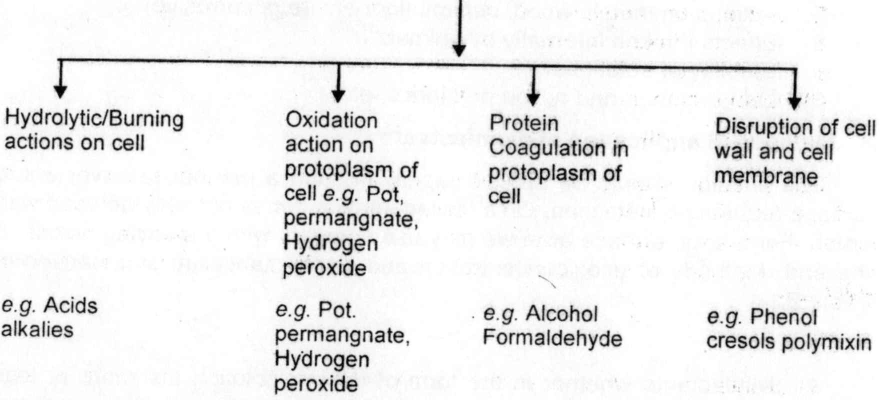

Factors Affecting Germicidal Action

1. Concentration of disinfectant.
2. Time of exposure.
3. Temperature of disinfectant solution.
4. Solution whether a fresh or stored.
5. Method of use–brushing, spraying, sprinkling or dusting etc.
6. Proper cleaning prior to disinfectant to bring germs in close contact.
7. Kind of surface, mud, metal, wood, rubber etc.

Four Essentials of Practial Work of Disinfection

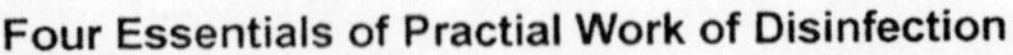

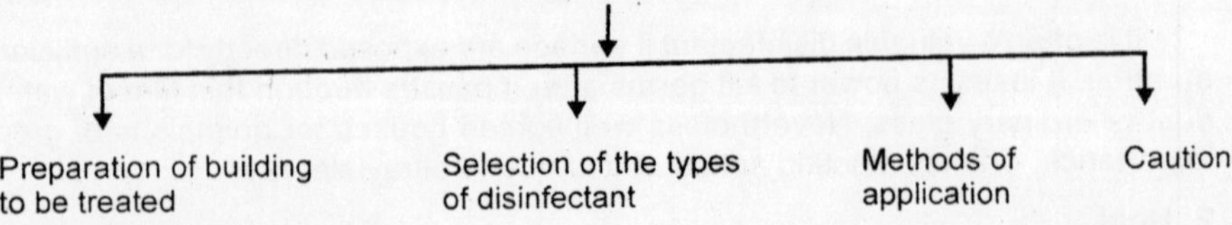

Preparation of building to be treated | Selection of the types of disinfectant | Methods of application | Caution

1. Preparation of building

The various surfaces, such as ceiling, walls, partitions, and floors should be swept free of cob webs, dust and dung. Any accumulation of filth must be removed by scraping, scrubbing with a wire brush, warm water.

2. Selection of disinfectant

All factors that enter into evalution of any disinfectant before a given compound is recommended are as follows :

1. Effectiveness–specific or general.
2. Properties of solubility.
3. Availability.
4. Cost.
5. Any additional preparation before use.
6. Toxicity of tissues.
7. Actions on metals wood, cement floor *etc.* (*e.g.* corrosive).
8. Effects if taken internally by animals.
9. Stability of solutions.
10. Odour, colour and action on fabrics, *etc.*

3. Methods of application of disinfectant :

The solution should be applied rapidly in such a manner to cover entire surface requiring disinfection. On a limited surface it may possibly be used with brush. For a large surface area we may use sprayers with a spraying nozzle at the end. Methods of use, concentration and disinfectants are summarised in Table 5.1

4. Cautions

All disinfectants whether in the form of dust or solution are more or less poisonous and irritating to eyes, skin and respiratory passages. Persons applying these must be careful to avoid ingention of these. Eyes, nose and mouth must be protected particularly from dust. Goggles, gloves, respirator *etc.* must be worn.

4.12 MAINTENANCE OF RECORDS

Purpose

1. Evaluation of animals.
2. Selection and culling of animals.
3. Systematic breeding programme.

Table 5.1, Common Disinfectants, Concentration and Method of Use

	Name (1)	*Concentration (2)*	*Method of use (3)*	*Approach Surface for use (4)*	*Remarks (5)*
1.	***Washing soda (Sodium Carbonate)***	3% soln. in boiling water.	Splashing Rinsing Utensils	Utensils & Floors	Little disinfection power but effective cleansing agent
2.	***Lime***	1/2 kg lime/gallon of water as white wash + 5% phenol	Sprinkling dusting of powder of lime	Floors, walls & grounds	Use freshly prepared solution.
3.	***Potassium Permanganate***	1:10,000 solution in water	Splashing	Floors, gutters & troughs	Disinfection action is due to oxidising power
4.	***Mercuric Chloride***	1 in 1000 parts of water	Splashing	Floors, gutters	Extremely poisonous, corronds metal, strong disinfectant.
5.	***Phenol (Carbolic acid)***	2 to 5% soln. in water	Splashing	Metallic objects & clothings	Good disinfectant
6.	***Formalin (40% formaldehyde)***	5% solution in water 15 to 20 ml formalin + 15 gm of Km O4 per m^2 space. 1% solution in water	Splashing & fumigation	Houses internal atmosphere	Door and window must be opened before housing pigs.
7.	***Silver nitrate***	1 % solution in water.	Dissolution sprinkling	Eye infections	Too expensive to use in large amount.
8.	***Copper sulphate***	28 gms in 8,000 gallons of water	Droops	Lakes, Prservoirs with ponds, plants	Useful to control algae & fungus growth
9.	***Lye solution (NaOH)***	2% for general use & 5% splashing for killing spores.	Splashing	Floors gutters equipments	Corrosive nature on metal, good, cheaper disinfectant

	Name (1)	Concentration (2)	Method of use (3)	Approach Surface for use (4)	Remarks (5)
10.	***Creosote***	2 to 4% hot solution	Splashing	Floor, gutters, woods troughs	Less toxic & cheaper
11.	***Ethyl alcohol***	70% soln. in water	Wiping	Skin, hands etc.	Protect from flame.
12.	***Iodine***	2 to 5 % solution in alcohol as tincture	Wiping	Skin, wounds, cuts etc.	Very effective poisonous & irritating
13.	***Bleaching powder (Calcium hypochlorite)***	30% available chlorine	Dusting	Floors, gutters, passages	Protect it from sunlight
14.	***Sodium hypochlorite***	200 ppm available Chlorine	Rinsing wiping	Utensils, udder	Don't use hot solution
15.	***Crystal violet***	2% solution	Wiping	wounds, skin cuts	Protect clothes from staining.
16.	***Quaternary ammonium salts (QAC)***	1% solution	Wiping/washing	Udders, hands	Bactericidal more effective against gram + ive than gram -ive bacteria.
17.	***Boric acid***	5-6% solution	Pplashing	Skin, floors, walls equipments, wounds	—
18.	***Sulphonamid***	—	Dusting powder	Wounds	effective against streptococi

4. Improvement of herd.
5. Economic feeding.
6. Detection of abnormal conditions of animals.
7. Fixing proper purchase/sale prices of pigs.
8. Better supervision and management.
9. To ascertain income and expenditure.
10. Determining cost of production.
11. To check the efficiency of labour.
12. To determine breeding efficiency.

Kind of records

1. Litter record
2. Individual pig register
3. Breeding register
4. Farrowing register
5. Daily feeding register
6. Health register
7. Ledger
8. Cash hook
9. Purchase book
10. Store stock hook
11. Bill payable book
12. Inventory register
13. Attendence/pay register
14. Weight register
15. Feed order book.

1. Individual pig register

Pig. No.	*Sex*	*Number of teats*	*Birth Weight*	*Coat Colour*	*Off colour marks*	*Weaning weight (kg.) of 56 days*	*Defects/ Remark*

2. Breeding register

Number of sows	*Date*		*Number of Farrowing completed*	*Date of last farrowing*	*Boar assigned*	*Date of service*	*Date of farrow*
	Birth	*First Farrowing*					

3. Farrowing record

Number of sow	Farrowing date	Number of piglets born	Number of pigs born			Total number of pigs raised
			Live		Dead	
			Male	Female		

4. Feeding register for the month for.....

Date	Feed amount (BF)	Feed Purchased	Total feed in stock	Number of pigs	Feed issued (kg.)	Balance of feed (CF)

Litter record

Litter No. — Breed —

Dam Record : Breed —

Number /Name................... [Sire
Dam

Birth Date....................

Number carcass.................... Average back fat ....cm

Sows...................litter (I, II, III etc.) Loin eye .... cm^2

Loin length.................... cm^2

Sire record : Breed —

Number/Name.................... Sire
Dam

Birth date....................

Number of carcss.................... Average back fat....................cm.

Loin eye cm2

Loin length....................cm

Health record :

Date Hog cholera vaccination..

Date Erysipelas vaccination ....

Piglets record : Number born (total) ....

Alive ....; Males ....Females..................

Date of deworming...

Mummies..................................Males.......................................Females..................

Pig weaned..............................Males ....Females..................

6. Inventory register

Sl. No.	*Animal No.*	*Sex*	*Class*	*Birth date*	*Age*	*Stock value*	*Depreciation*	*Present value*

5.13 CARE OF ANIMALS

5.13.1. Care of piglets at birth

1. Remove the piglets soon after they are farrowed.
2. Clean all piglets and make their body dry.
3. Make breathing passages of all baby pigs clear.
4. Cut the navel cord with sterilized scissors leaving 3 cm from navel and disinfect by application of tincture of iodine.
5. Allow piglet of suckle milk from mother sow for about 8 to 10 times in 24 hours initially.
6. Protect the baby pigs from trampling by sow.

Feeding of piglets

Age	*Daily feed*	*Cumulative feed in kg.*
Birth to 1 month	Mother Milk	–
1 to 2 months	0.5 kg	15
2 to 3 months	1.0	30
3 to 4 months	1.2	36
4 to 5 months	2.0	60
5 to 6 months	2.5	75
Total		216

Note : About 216 kg of total feed is expected to produce a pig of 70 kg. body weight in 6 months.

Prevention of anaemia : (Nutritional anaemia in suckling pigs).

It is a highly fatal diseases of suckling pigs caused by a marked decrease in haemoglobin and fatty degeneration of the liver.

Causes : Lack of iron and copper salts in milk of sow pigs kept in indoor pens, on concrete floor and limited milk diet from sow.

Age group of pigs affected : 3 to 6 weeks.

Symptoms : Pigs are dull and inactive, fatigued, lack of vigour. Pig show dyspnea, thumps and rough coats, depression on the slight exertion. Pig may be weak and thin, fat yet the muscles are flabby. Skin over the neck and mucus membrane are pale. Pigs die suddenly. Wrinkles are found over the legs. Pigs may develop diarrhoea.

Prevention and treatment

1. Add small amount of iron and copper in pigs diet @ 25 mg iron and 5 mg of copper/day/pig.

2. Iron sulphate – 3.6 ounces
   Water – 5 quarts

Note : Feed 1 dram daily

OR

3. Paint the udder of sow daily with the following mixture :

Iron sulphate	–	500 g.
Copper sulphate	–	70 g.
Sugar	–	500 g.
Water	–	10 lit.

OR

4. Allow piglets to free access for parasite free runs with fresh oil.

OR

5. Intra-muscular injection of iron–dextrose compound.

Creep Feeding : Start feeding piglet creep feed from 2 to 3 weeks age for proper growth and development. Area for creep feed should be partitioned so that all piglets may have access to it. See Table 5.4 for creep feed composition.

Note : Creep feeding has a pronounced effect on livability as there is a significant difference (P 0.01) in mortality between those creep fed and others (Rao and Rao 1981).

Removal of needle teeth : Baby pigs at birth have 4 pairs of sharp teeth on each jaw called needle teeth. These are not of any use to piglets and may cause injury to mothers udder. These should be clipped by means of plier. Care be taken not to injūre jaw or gum of piglet.

Raising orphan piglets

Causes of orphan piglets :
1. Large size litters than a sow can raise.
2. Death of sow after farrowing.
3. Failure of lactation.
4. Mastitis.

Method of raising

1. Use of foster sow
2. Use of milk replacer

Egg york –1
Cow Milk –1 kg] Mix thoroughly

Note : To see that foster sow accepts new piglets these must be sprayed with some deodorant disinfectant to maks the smell.

Weaning Piglets

1. Wean the piglets at 8 weeks of age.
2. For weaning separate the sow from piglets for few hours everyday.
3. Supply creep feed with 18 per cent crude protein.
4. Deworm the weaned piglets at 10 week age.

Causes of preweaning mortality

Almost 61.33 per cent of the deaths of piglets occurs in the first week. Hence good husbandry practices in early life will be profitable. The major causes of deaths were found to be agalactia, pnemonai and gastroenteritis. Strict hygeinic measures may help in reducing the mortality due to infectious diseases. Piglets with birth weight were prone to agalactia, trampling and infectious diseases. Therefore, special attention to pregnant sow nutrition may reduce preweaning mortality.

5.13.2. CASTRATION OF PIGLETS

5.13.3 Care and Management of Young Stock

Birth weight of the piglets and weekly body weight should be recorded to watch for feed efficiency and also for culling and disposal of uneconomic animals. Wean the piglets at 8 weeks age and raise them in small batches. Deworming can be done by use of suitable anthelmintics. Ectoparasites if present can be removed by external application of 0.5 per cent sumithion. Good virile piglet must be raised as boars and other should be castrated at 6 to 8 weeks of age. These animals are allowed to fatten well. Usually at 22 to 26 weeks age pigs will attain a live weight of 60 to 80 kg and these are ready for disposal for slaughter.

5.13.4. Care of Management of Gilts and Sows

1. Maturity of gilts is more important than its age. Larger litter usually results from gilts bred during their third heat period.
2. It is better to purchase a gilt that has been already bred alteast once. Although it is advantageous to keep a brood sow. Brood sow are at peak of productivity during 2 to 3 years age.
3. Flush the gilt or sows (feeding liberally) to make sure a gain of about 500 g per day during **two** weeks before breeding. It helps to increase sheding of more number of ova and thus increase litter size. For this feeding of 3 to 4 kg. good ration with 16 per cent crude protein and balanced with mineral, vitamins is adequate.
4. Heat period in sows lasts for 2 to 3 days, these should be mated 2nd day of heat, preferably 2 services at 12–24 hours interval. Female if not bred will repeat heat at 21 days interval.
5. Record the dates of mating, expected farrowing, *etc.* keeping in view the gestation period is 114 days.
6. Provide balanced ration containing 12 to 14 per cent protein, 0.8 per cent calcium, 0.6 per cent phosphorus to the pregnant females.
7. Leguminous fodders like berseem, lucerne, cowpea must be made freely available.
8. The ration for breeding stock must have 16 per cent protein will balanced with cereals, cakes, by–products, minerals and vitamins in proper ratio and amount.
9. Ration amount/day per 45 kg of live weight :

   Gilts–1.25 to 1.5 kg

   Sows–1 to 1.25
10. All gilts and sows be maintained at a lower temperature because the survivability of embryo is higher at a low temperature (16°C) than when maintained at 32°C temperature.
11. Keep houses clean with controlled environmental conditions to avoid stress on animals
12. Periodic spraying of insecticides in houses is essential to keep free of ectoparasites and protect them from mange.

13. Arrange separate enclosure for breeding purposes and little away from boar sty.
14. Arrange a stall to keep sick animals separately.
15. Inbreeding a herd should be avoided to prevent decrease in fertility.

5.13.5 Care and Management of Sows and Gilts in Lactation

1. Do not give full amount of feed to sow recently farrowed.
2. Increase the amount of feed gradually every day.
3. Take at least one full week to get the sow on full feed, otherwise piglets may have scours.
4. Amount of feed for sow per day be determined @ 500 g per piglet in her litter.
5. Crude protein per cent and nutritive ratio in the ration of nurse sow be 15 and 1 : 15 respectively.
6. Composition of ration for nurse now (See Table 5.4).

5.13.6. Care and Management of Pregnant Animals

1. Record the date of mating to determine expected date of farrowing because swine have a short gestation period to 112 to 116 days.
2. Provide a separate sty of about *3m²* for each sow.
3. Keep the sty clean, dry and nonslippery.
4. Keep her liberally fed and kindly treated.
5. Do not allow to mix pregnant gilts/sows with other animals to avoid fighting which may lead to skidding and abortions.
6. Avoid rough treatment–kicking, abusing, pushing *etc.* when spoils the temperament.
7. Do not allow her to walk for a long distances to avoid fatigue.
7. Offer clean, safe odourless water at all times in sty.
8. Provide a bedding of 8 to 10 cm chopped straw under covered area.
9. Allow pregnant sow to graze on pasture with the following conditions :
   (a) Separate pasture from other animals.
   (b) Safe from parasites.
   (c) Nutritious and green grass.
   (d) Grazing in the morning hours only.
   (e) Refer for more details (pasture for hogs under feeding section).

   or

   Provide balanced ration with 16% crude protein and 1 : 5 nutritive ratio, 0.8 per cent Ca, and 0.6 per cent P for uniform litter with more birth weight. Allow free access to leguminous greens like berseem, lucerne, cowpea.
11. Increase the amount of feed during last month of gestation because major growth of foetus takes place in this period.
12. Deworm the pregnant females 2 weeks before farrowing with piprezine adepate @ 1 gm/10 kg body weight.

Note : Danielson *et. al* (1991) reported that sows dewormed before farrowing consumed less feed during lactation even though they nursed and weaned more pigs than control group. Total litter weight was greater for litters from dewormed sows.

13. Give bath to pregnant females with luke warm water containing pot. permanganate (1 : 1000) or allow free access to medicated wallow, only after they have been watered.

;.13.7. Farrowing Care and Management

)bjectives

1. To reduce mortality of piglets.
2. To ensure better lactation.
3. To maintain high level of breeding efficiency.
4. To ensure well being of both sow and piglets.

)esirable conditions in farrowing pen

1. Cleanliness and good hygiene.
2. Easy to supervise farrowing in gilts and sows.
3. Provision of suitable farrowing crates to prevent sow from flopping directly from standing position.
4. Provision of creep area of safe zone in the pen.
5. Arrangement of light and heat to protect piglets.
6. Provision of guard rails to prevent crushing losses of piglets.
7. Proper arrangement of feeds and water.
8. Adequate floor area.

Management before farrowing

1. Trim over grown toes to the sow to minimise crushing injuries to piglets.
2. Clean and disinfect farrowing pen to prevent infection.
3. Washing, cleaning and disinfection of sow just before transfer to farrowing pen.
4. Provide beeding of wheat or paddy straw on the floor in farrowing stall to provide warmth to the sow and newly born piglets.
5. Transfer the sow to farrowing pen about 8 to 10 days before farrowing date to make her accustomed to her new environment.
6. Provide bulky ration in farrowing pen.
7. Reduce the ration by 1/3 till farrowing.
8. Withdraw feed 12 hours before farrowing.
9. Provide free access to fresh water.
10. Keep floor dry and sow confortable at a temperature below 24°C.
11. Appoint attendent/experienced hand to watch and look after the sow.

Signs approaching farrowing

1. Restlessness and abdominal contractions.
2. Attempts to make nest from bedding material, the sow tends to paw the floor.
3. Teats become prominent with a change in texture of udder.
4. Teats are filled with colostrum.
5. Noise level increase due to discomfort and labour pain.
6. Chewing any object in pen due to nervousness and frustration.
7. Expulsion of the blood stained fluid from vulva.
8. Slight increase in rectal temperature by 1°F.
9. Expulsion of meconium from the vulva.
10. Twitching of tail.
11. Drinking water, urination and defectation more frequently.

Stages of farrowing

Stage I. Dilation of phase of cervix
Stage II. Phase of foetus expulsion.
Stage III. Phase of placenta expulsion.

Period of complete farrowing : 2 to 4 hours with an average interval of 12 to 20 minutes between birth of piglets.

Note : The interval between birth is longer in older sows and shorter in gilts due to better muscular tone in gilts.

Disposal of placenta

All the placentae are thrown one after another within two hours after farrowing. This must be collected manually and buried for disposal. Prevent sow eating her placenta.

Presentation and position of fetuses in pigs

The relationship between the longitudinal axis of the fetus with that of dam and dorsum of the fetus to the maternal pelvic quadrants at birth where defined as presentation and position of the fetuses, respectively. Solmonaj (1991) reported following types of presentation of fetuses in indigeneous pigs at birth and in the birth order :

Presentation type	*First*	*Middle*	*Last*	*total*
	Third	Third	Third	
Anterior %	65.91	48.78	55.17	57.02
Posterior %	34.09	51.22	44.83	42.98

Position of the fetuses at birth

Dorsal–Sacral	84.74 per cent
Dorso–Paleral	4.38 per cent
Dorso–Pubic	0.88 per cent

Care and Management during farrowing

(i) Care of baby pigs just after birth

1. Arrange experienced person to attend, irrespective of time of day or night.
2. See that all breathing passage of piglets are made free. In case of difficulty swing the piglets violently round at arms length or immersed in a container of cold water to stimulate piglet to gasp for breath and initiate breathing.
3. Make piglets free of mucus and dry.
4. Transfer piglets to creep area in which temperature is 25 to 30°C, to protect from chilly weather.
5. Prevent sow from eating her placenta and dispose off the placentae.
6. Save the piglets from nervous/irritable acts of sow.
7. Assist piglets to get the teat and suckle colastrum from their mother's udder.
8. Check the udder for agalactia (lack of milk), mastitis (inflammation of udder) and number of functional teats to assess the nourishing capacity of sow.
9. Cut the navel cord with sterilized scissors leaving 2.5 cm from the body and disinfect with iodine solution.
10. Provide veterinary aid in case of abnormal/difficult farrowing.

Note : If farrowing is manually essential, give injection of antibiotics.

11. Provide micro–environment of light and heat to keep piglets comfortable from inclement weather.
12. Provide supplementary feeding incase inadequate nourishing ability of sow.

(ii) Care of sow just after farrowing.

1. Clean the sow with luke warm water and allow her to suckle her young ones.
2. Gives oxytocin injection if there is no let down of milk.
3. Give sow her first meal 12 hours after farrowing.
4. Giver her free acess to clean, fresh and safe water.
5. Check rectal temprature which if above 105°F then give appropriate treatment. (normal body temperature of sow is 103°F just after farrowing).
6. "Even up" the litters by cross fostering of the same sized piglets.
7. Spray the sow's udder with saturated solution of ferrous sulphate (50 g of $FeSO_4$ in 1 litre of water), to prevent anemia.
8. Provide adequate nutritious balanced ration to the lactating sow for sufficient milk production for the piglets. (Refer 5.4.)

5.13.8. Care and Management of Boar (Sapra and Mehta, 1990)

Selection

1. Select the pure bred boar and good representative of the breed it belongs to.
2. Look for pedigree and conformations.
3. Boar influences conception rate, litter size and contribute one half of the genetic make–up of his daughters. Hence boar must have well developed sex organs.
4. Boar must have signs of masculanity.
5. Boar must possess the traits of meaty hog. Soundness of feet, legs, underline, body capacity should be observed. Boar lacking sex derive must be culled.
6. Never select a boar that has one testis (cryptorchid).

Age for breeding and use

1. Suitable age for breeding–10 to 12 months.
2. New boar for replacements must not be used until after a minimum of 40 to 60 days isolation period. This permits a health check and also good adjustment with surroundings.
3. Use one boar for 10 to 15 sows or 7 to 8 gilts per week.
4. Number of services per boar per week should be 4 to 5.
5. Use one boar to set of boars one day and another boar or set of boars next day.
6. Bring boar to sow or sow to the boar at the time of service rather than allow the boar to run with female stock all the time. A young boar turned with large number of sows may become shy.
7. For optimum breeding performance do not use young boar for more than one service daily and mature boar for not more than 2 services a day.
8. Boar will give service satisfactorily upto 5 to 6 years age.
9. First service is critical in the life of a boar. Therefore this mating should be with a female of his own age and size.
10. Make use of breeding crate for mating purposes. Avoid slippery floors in the breeding areas.

Feeding of boar

1. Boar should neither be underfed nor fattened but kept in food thrifty condition.
2. amount of feed per 45 kg body weight per day should be 1.25 to 1.50 kg.
3. The ration (see Table 5.4) should contain 14 per cent protein with a nutritive ratio of 1 : 5.

4. Boars should be fed 2 to 3 hours before they are expected to breed.
5. Provide fresh, clean and adequate quantity of water.

Housing

1. Boar pen should be located where it will provide maximum exposure to the females to be bred.
2. The precence of females helps to stimulate the sex desire in boars.
3. Each boar should be provided with 1.5 to 2m^2 of dry draft free, well ventilateed sleeping area.
4. Hogs have very few sweat glands and hence cannot dispose off the body heat as easily as other mammals do. Therefore, to protect from heat stress/heat induced sterility provide bathing facility of wallow specially in summer.

Temperature

Temperature affects behaviour, rate of gain, feed utilization carcass composition, nutritive requirement and reproductive efficiency.

Note :

1. An irritable boar difficult to drive or inclined to fight may transmit a nervous disposition to offsprings which can make them poor mothers.
2. Periodically carry out a fertility check by observing heat repeaters. If one out of 5 gilts returns to heat within 30 days, it reflects the low fertility of boar.

Brief account of good management practices

1. Protect feed and water from contamination of manure, urine, etc.
2. Provide good drainage to prevent mud puddle, and keep area clean.
3. Carry out regular deworming of herd.
4. Periodical spray of insecticides to protect from lice and mange, etc.
5. Keep different class of pigs young, old, weaners, breeders, pregnant stock, boar, separately.
6. Provide comfortable accomodation, dry, clean, ventilated and draft free shady area.
7. Provide iron to baby pigs to prevent nutritional anaemia.
8. Feed well balanced diet low in fibre.
9. Provide access to clean, green and parasite free pasture or else provide liberally enough green succulent fodders like lucerne, berseem, cowpea.
10. Segregate sick animals.
11. Vaccinate hogs for swine fever and hog cholera.
12. Follow a good calender of daily operations.

5.14. PREPARING HOGS FOR SHOW OR FAIRS

Purpose

1. To exhibit best type and win the contest.
2. Enables breeders to exchange experiences.
3. Provides opportunity to make comparisons among superior types both within and between breeds.
4. Permits new breeders to make contacts with established breeders.
5. Serves to popularise best genetic make–up of breeds.
6. Serves best advertising media.
7. Increase pride of farmers.
8. Opportunity to learn more of good feeding and management.
9. Arouse interest in other farmers.
10. Builds a healthy competitive spirit.
11. Welfare a people and hog industry.

Materials

1. Good type of hog (true to the breed)
2. Brush
3. Soap
4. Pig catcher
5. Coconut oil/paraffine oil
6. Hoof trimmer and rasp
7. Warm water
8. Corn starch
9. Hair clipper
10. Pair of scissors
11. Buckets
12. Strong rope
13. Dusters
14. Blue Dye
15. Bolt Clipper
16. Cane

Procedure

Steps	*Reasons for*	*Precedural details*
1. Selection	To complete	Choose the best type with true breed characters, good temperament, free from defects, good growth as per its age and optimum condition of fleshing , two months before show.
2. Gentling/ Taming	For judges to see hog and for close observations, with advantages.	Tame the pig well in advance using cane for guiding and training for walk and stop when desired at will. Make it so gentle that it may be guided by just showing the cane or placing it along the side of head.
3. Trimming toes	To make hog stand squarely and walk properly.	Push the body of hog and make it lie on one side, or else use pig catcher to keep under control. Trim the toe by hoof trimmer and bring into proper shape and length. Use hoof rasp to square-up the soles just three to four weeks before show.
4. Removal of tusk	For sale handling and to prevent injury to other animals.	Cast the pig and tie to a post. Raise the upper jaw with pig catcher and remove the tusk by use of bolt clipper.

Steps	Reason for	Precedural details
5. ***Clipping hair***	For increasing external appearance, trimness and show off teats clear.	Few days before the show the hair on tail except the switch, long hair over the hocks, body and belly of gilt must be removed by use of clippers.
6. ***Bathing***	To make body clean, skin smooth and mellow.	Make the body wet with water. Apply soap and rub. Wash the body. Keep the animal in a clean sty.
7. ***Oiling***	To make skin soft, smooth and give a bloom to hair coat.	Soak the small piece of cotton in oil and apply over the body and keep in clean place. A further light application is desirable just before the show and light rubbing of body before entering the ring.
8. ***Powdering***	To make hogs clean and attractive.	Wash the pig and make dry. Just before entering the ring powder the body with corn starch to make white spots on dark hogs very clear and white pigs clean and attractive.
9. ***Showing***	To win the contest.	**Use following guidelines :** 1. Tame and train the animal long before the walk stand and pose as desired. 2. Keep the animal ready well groomed and watered. 3. Make animal to walk before judges. 4. Dress up nicely for this occassion. 5.Enter the pig immediately when called. 6. Keep the animal in a good pose before the judges at a distance. 7. Keep a watch on animal. 8. Keep calm with no sign of nervousness. 9. Prevent animal from bite and fight with other hogs. 10. Be humble and courteous to visitors too.

4.15. TRANSPORTATION OF PIGS

The quality of the carcass is adversely affected by handling methods of animals on the farm, during transport and the slaughter. Many different factors cause stress of fatigue during transport are separation from a familiar environment and social regrouping, the process of loading and unloading, climatic conditions, design of loading equipment, loading density, duration of transport, driving technique, vibrations and noise.

All these factors have an effect on the condition of the animals on arrival at the slaughter house which affect post mortem meat quality. Under transport conditions, pigs are often in the upper level of thermal tolerance and a small increase in environmental temperature will not have an effect on pig. But in combination with forced confrontation and strange conspecities, high stocking densities and poor design of loadign equipment may cause fatility. The number of factors crucial for condition of pigs for obtaining quality carcass are as follows :

Different modes of transporation

* By road (truck)
* By rail
* By air
* By ship.

Transport by truck is the most common way of hauling pigs. Ships and train transports are seldom used and air transport is too expensive for slaughter animals, but is used for transporting breeding pigs. Though road transport is worse for the animals than rail, sea or air transport, it is more economical and easily accessible.

Transport by Road

Transport of pig by road has the convenience of loading at the farm and direct transit to the destination, the absence of repeated handling and the disturbances associated with it. Yet, during road transport factors such as weather conditions (air temperature, velocity, and humidity), loading density and duration of trip influence the condition of the pig on long trips of two to three days or more, during which pigs are exposed to wide variations in weather conditions. In general, live weight losses during transport of 1 to 2 days are 40 to 60 g per day whereas the mortality is 0.1 to 0.4 per cent. Transport conditions for pigs can be improved by using loading ramps with a slope of less than 200 and lighting the inside of the truck.

Feeding and Watering

Feed should be withheld on the day pigs are transported and water should be available until the moment of loading, Death losses during transport were lower when feed was withdrawn before transport. It has been suggested that the pigs die by choking and by inhalation of their own vomit. Vomiting was sometimes observed during transport which may be caused by food intake shortly before loading.

In hot climate areas, pigs should be transported at night or before 10 am in the morning. High temperatures during transport have a nagative effect on animal welfare. Therefore additional cooling may be beneficial. The lowest heat production and best meat quality occur at an environmental temperature of 16°C and an air velocity of 0.2 m per second. Wetting pigs prior to transport and showering on the truck reduced the effect of heat stress and death losses. Deep bedding of straw will protect pigs from extreme cold. Straw should not be used during warm weather because it will cause the pigs to become over heated. Pigs should have access to water and one–hour rest perioa for every 8 hours of journey, and pigs should not be transported for longer than 24 hours. They should be rested, fed and watered for 10 hours after a 24-hour journey. During the resting period, all animals must have sufficient place to lie down either on or off the trucks.

Loading density

Loading density has major effect on animal welfare and post mortem meat quality. Tight stocking ($0.3m^2$ per 100 kg pig) results in more skin blemishes and rectal prolapse. A loading density of approximately 285 kg per m^2 is suggested to be an acceptable compromise between animal welfare, meat quality and economics of transport.

Recommended loading densities for pigs during transport (all animals are able to lie down at the same time at these densities) :

25	0.15	165
60	0.35	168
80	0.40	200
100	0.42	235
120	0.51	235
140	0.51	235

Fasting and meat quality

The percentage of PSE (pole soft exudate) meat is lower if pigs are fasted before loading. There are other advantages of fasting prior to transport, such as less labour at slaughter, less contamination of the carcasses, less spillage of gut contents, and a decreased weight loss of the carcass during chilling. A period of feeds withdrawal of 12–24 hours before slaughter is recommended to improve meat quality, while the losses in carcass weight is minimised. However, fasting for more than 24 hours may increase the percentage of DFD (Dark Firm Dry) meat in the carcass. On long trips where pigs will be off feed for more than 24 hours, feeding a thin porridge prior to transport will reduce the live weight losses. The porridge consists of one part of feed and three parts water. Under normal housing conditions a pig drinks 7 to 20 litres of water per day. Stress during transport may reduce the intake of water.

6

Health and Hygiene for Hogs

6.1 CLEANING AND DISINFECTION

Refer details in previous chapter (Sanitation and disinfection)

6.2 DISEASES OF SWINE

Pigs are susceptible to many diseases. However most of these diseases can be prevented by timely care, sanitation and hygiene. Prevention is highly effective and also economical.

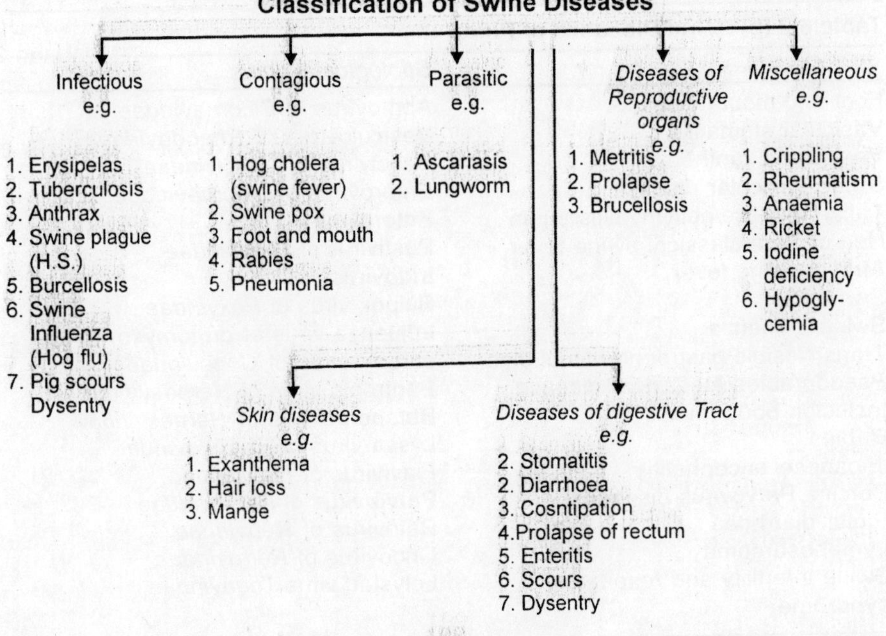

Table 6.2 (a) : Bacterial Diseases of Pig

Diseases	*Etiological A gents*
Swine erysipelas	Erysipelothrix rhusiopathiae
Glassers pig diseases	Haemophilus suis
greasy pig diseaes	Staphylococcus hyicus
Cytitis and pyelonephritis of sows	Corynebacterium suis
Brucellosis	Brucella suis
Foot rot	Spherophorus necrophorus
Listeriosis	Listeria monocytogenes
Leptospirosis	Leptospira pomona, *L.* icterohaemorrhagiae
Tuberculosis	Mycobacterium bovix, *M.* tuberculosis
Anthrax	Bacillus anthracis
Pasteurellosis	Pasteurella multocida
Colibacillosis	Enterotoxigenic E. coli
Bordetellosis	Bordetella bronchoseptica
Tetanus	Clostridium tetani
Infectious necrotic hepatitis	Clostridium novyi
Enterotoxemia	Clostridium perfringens
Salmonellosis	Salmonella choleraesuis
Actinomycosis	Actinomyces bovis
Enzootic Pneumonia	Mycoplasma hyponeumoniae

Table 6.2 (b) : Viral Diseases of Pigs

Viral Diseases	*Etiological A gents*
Foot and mouth disease	Apthovirus of *Picornaviridae*
Vasicular stomatitis	Vesiculovirus of *Rhabdoviridae*
Vesicular Exanthema	Calicivirus of *Caliciviridae*
Swine vesicular disease	Enterovirus of *picornaviridae*
Taflan disease/pollymyelitis suum	Enterovirus
Hog cholera/classical swine fever	Pestivirus of *Togaviridae*
African swine fever	Iridovirus
Swine Pox	Suipox virus of *Poxviridae*
Swine Influenza	Influenza virus of *orthomyxoviridage*
Transmissible gastroenteritis of pigs	corona virus of *Conoviridae*
Pseudorabies/Aujeszkys disease	Pantropic virus of *Herpesviridae*
Inclusion body rhinitis	Betaherpevirus of *Herpesviridae*
Rabies	Lyssa virus of *Rhabdoviridae*
Japanese encephalitis	Flavivirus of *Flaviviridae*
Porcine Parvovirus diseases	Parvo virus of *Parvoviridae*
Piglet diarrhoea	Ratovirus of *Reoviridae*
Lymphosarcoma	Oncovirus of *Retroviridae*
Swine infertility and respiratory syndrome	Lelystad virus *Togaviridae*

Table 6.2 (c) : Metabolic and Mineral Deficiency diseases

Diseases	*Mineral Deficiency*
Piglet anaemia	Iron
Parakeratosis	Zinc
Parturient paresis	Calcium
Rickets	Calcium, Phosphorus, Vitamin D
Osteomalacia	Calcium, Phosphorus, Vitamin D
Atrophic rhinitis	Calcium, Vitamin D
Ketosis	Negative energy balance
Neonatal hypoglycemia	Inadequate intake of milk
Enzootic ataxia	Copper
Myxedema	Iodine
Mulberry heart disease	Selenium/Vitamin E
Nutritional muscular dystrophy	Selenium/Vitamin E
Hepatitis dietetica	Selenium/Vitamin E
Exudative diathesis	Selenium/Vitamin E

Table 6.2 (d) : Parasitic Disease of Pigs

Parasites	*Common Names*
Fasciola sp.	Liver fluke
Fasciolopsis buski	Intestinal fluke
Gastrodiscoides hominis	Conical fluke
Diphyllobothrium latum	Tape worm
Cysticerus cellulasae	Pork measles
Ascaris lumbriwides, A. suum	Large round worm
Strongyloides ransomi	Intestinal threadworm
Metastrongylus april, M. salmi	Lung worm
Hyostrongylus x rubidus	Stomach worm
Trichinella spiralis	Stomach worm
Trichuris suis	Whip worm
Stephanurus dentatus	Kidney worm
Dioctophyma renale	Large Kidney worm
Oesophagostomum dentatum	Nodular worm
Macracanthorynchus hirudinaceus	Thorny headed worm
Ascarops strongylina	Spirurid worm
Gongylonema pulchrum x	Spirurid worm
Physocephalus sexalatus	Spirutrid worm
Elimeria perminuta E. licabra E. spinosa,	Coccidoiosis
Isospora suis	Coccidiosis

6.3 DETAILS OF SOME IMPORTANT SWINE DISEASE AND PARASITES

6.3.1 Hog Cholera (Swine Fever/Pig Typhoid)

It is highly contagious and usually acute. Young hogs are more susceptible than older ones. It is an important exotic disease entered in India in near 1960 and caused severe loss of pig population.

Cause

Filtrable virus which usually gets into the animal's body through digestive tract, from which it enters the blood stream and distributed throughout the body.

Transmission

1. Through urine and faeces, nose as well as mouth secretions.
2. Contaminated feed and water.
3. Addition of new animals into the herd or by shoe of attendants.
4. By bird or streams
5. Direct or indirect contact with affected pigs, dead or alive.

Note : Virus is even present in the urine of healthy pigs after serum virus inoculation. Therefore, recently inoculated pigs may spread disease to susceptible animals.

Incubation period : 7 to 10 days.

Symptoms

Fever, loss of appetite, dullness, dejected appearance. Body temp. 40°C (106°F)–later it fails to normal or below normal, low intake of food, weakness, staggering gait (wobbling), eye discharge often present so eye lids stick togeather, initially constipation but diarrhoea often follows. Red bloches (which do not blanch when pressed) may be seen on the skin–especially on the abdomen. Infected animals cough and have difficulty in breathing. Usually animal dies within seven days and some who survive show partial recovery.

Prevention

1. Proper sanitation, proper feeding and good care.
2. Old litter other material or contaminated premises should be burned.
3. Houses and equipment thoroughly cleaned and disinfected.
4. Administration of anti–Hog Cholera serum to suspected cases.
5. Vaccination of healthy pigs and serum virus inoculation or lapinised swine fever vaccine.

Treatments

No effective.

6.3.2 Swine Para Typhoids

Cause

Bacteria – Salmonella choleraesuis var. ***Kunzendorf.***
Infectious bacterial disease occuring in acute or chronic form.
Other organisms which affect G.I. tract *S. typhimurium ; S. typhsuis*

Symptoms

Fever, anorexia and red/purple discolouration of skin on certain parts of the body followed by diarrhoea, faeces with shredes of necrotic debris.

Treatment

Same as for swine fever.

6.3.3 Swine Dysentry

(Bloody diarrhoea, bloody scours, black scours, colitis)

Causes

Bacteria : *Hamorrhagic enteritis, swine typhus*

Symptoms

Profuse blood diarrhoea, off the feed, Temp. 103 to 104°F, illness linger for two weeks, unthrifly and those recover are stunted. Death.

Treatment

Use of sulphamezathene, phthalysulphathiazole, Streptomycin, Auromycin.

Prevention

Sanitation, quarantine.

6.3.4. Swine Vesicular Exanthema

Cause

Fltierable virus of 13 to 20 mili–micron size.

Symptoms

Infectious, formation of vesicles on the snout, mouth cavity, feet and other parts like nose, gums, tongue, dew claws, ball of foot, top of hoof, udder in nursing sows.

Source : Garbage feed.

Note : It may be mistaken as foot and mouth and vesicular stomatitis.

Prevention

Quarantine, sanitation.

6.3.5. Swin Erysiplas

Cause

Bacterium : *Erysipelothrix rhuslopathiae*

Symptoms

Stiff joints, lameness, sluggish and easily fatiqued. Hogs resting on their hounches or their breast bone.

Appears in three recognizable forms : the acute or septicaemic form, skine form and chronic form. Skin form is characterized by diamond shape utriearial lesions while vegetative endocarditis and arthritis are features in chronic case.

Transmission

Affected animals pass off causal organism of this in urine and faeces, hogs pick it up from contaminated food and water. Organisms remain alive and multiply in moist soil. In India, it is common from June to Sept. Mortality upto 50 per cent.

Prevention

1. Segregation of sick animals.
2. Proper balanced nutrition.
3. Administer protective dose of antiserum, injection be given behind the ears and not thighs along with the Pencillin.
4. Vaccination for longer and active immunity in April and May.

6.3.6. Swine Pox

Cause

Filtrable suipox virus, member of pox viridae family.

Transmission

Direct/indirect, through air, feed water and contact, etc.

Contigious skin disease.

Symptoms

Fever, lack of appetite, eruptions, over the skin commonly on ears, under surface of body and thighs. In severity lesions on all parts of skin, ulceration in digestive tract.

Prevention

1. Segregation.
2. Supply light diet.
3. Slaughter of affected animals.
4. Good sanitation, strict hygiene, clean and dry quarters.

Note : An attack of the disease makes the animal immune to rest of the life. Vaccination is not preferred. Disease can be eradicated by epidemiological methods of control.

6.3.7. Swine Plague (*Haemorrhagic septicaemia*)

Cause

Pasteurella suiseptica (bipolar bacterium).

Transmission

Being infectious it may occur by indirect means, contaminated feed, water infected litter and also by direct contact.

Symptoms

High fever, loss of appetite, staggering gait, difficult respiration, oedematous swelling of the throat, diarrhoea/disentry.

Prevention

Segregation, disinfection of premises, inoculation of animals with anti *Haemorrhagic septicaemia* (H.S.) serum, proper disposal of carcass and infected litter by burning deep, prophylactic measure of vaccination with H.S. Vaccine.

Treatment

Administration of broad spectrum antibiotics and injection of Dexona.

6.3.8. Foot and Mouth Disease

It is most contagious disease of all cloven footed animals.

Cause

Filtrable virus three distinct strains of family picornviridae ; major types are virus– O, A, C, SAT–I, SAT–II, SAT–III and Asia–1. In India, O.A.C. and Asia. 1 are common.

Symptoms

High fever, appearance of vesicles followed by ulcerations on mucus membranes of the mouth, foot lesions are painful, pigs can hardly walk. lie down for long time, indisposed to eat their food, usually not fatal but mortality may occur in young ones.

Prevention

Isolation, proper disinfection of premises, rubbing and alum on the lesions of mouth, treatment of lesions by $KMnO_4$ solution and sulphanelamide powder mixed with iodoform, vaccination.

Treatment

No effective.

6.3.9. Brucellosis (Contagious Abortion)

Cause

Brucella suis.

Symptoms

Premature abortion of litter, sterility, inflammation of uterus, testes and joints.

Prevention

1. Testing of all animals and reactors be marked and sold.
2. Purchase of stock from brucellosis free herd.
3. Isolation for at least five weeks after purchase of new stock.
4. Proper disinfection of infected litter.

6.3.10. Swine Influenza (Hog Flu)

Cause

Hemophilus influenza suis (bacterium).

Symptoms

High fever, loss of appetite, cough and discharge from eyes and nose, sluggish movement, low mortality.

Prevention

1. Provide warm, dry clean quarters.
2. Correct feeding and management.
3. Expectorants aid if respiration is difficult.
4. Rotation of pastures.
5. Prevent chilly drafts on animals.

6.3.11. Hypoglycemia (Bady Pig shakes)

Cause

Low blood sugar level.

Symptoms

Shievering, weakness, failure to nurse, pigs if disturbed emit weak crying squeel. Hair becomes errect and rough, death in 24 hours occur mostly in baby pigs.

Treatment

1. Give injection of dextrose solution intraperitoneal.
2. Force feeding of frequent intervals of a mixture of 1 part corn syrup with 2 parts of water.

6.3.12. Anaemia (Nutritional)

Cause

Iron deficiency, may also be caused by copper, cobalt and or deficiency of vitamin B.

Symptoms

Loss of appetite, amaciation and death, laboured breathing, swollen condition around the head and shoulders. Mostly prevalent in suckling young; suckling pigs lick concrete or wood if they have no access to soil.

Treatment

1. Provide dietary source of nutrients (ferrous sulphate)
2. Swab sow's udder daily with solution of half kg ferrous sulphate and 100 gm of honey in half kg of water.

6.3.13. Haemolytic Disease of Newborn Piglets

Clinical Signs

Signs occur 1 to 4 days after birth and consist of pallor, in activity, dyspnea and jaundice. These signs gradually subside. After 10 days, multiple petechial haemorrhages occur the ventral areas of the body; most piglets die at this time RBC drops from the normal at 4 days of age.

Diagnosis

It is amde on finding anaemia, purpura, and mild icterus is neonatal pigs.

Treatment and Prophylaxis

A substitute dam may be tried. Sows should not be rebred to the same boar after having effected litters.

6.3.14. Swine Parasites

(i) Ascariasis

Cause : Round worms/ascarias.

The round worm is large, thick, yellow or pink–lives in the intestine. Thousands of eggs are produced by femals daily and eliminated with droppings.

Infestation : Pigs get infested by swallowing the eggs with feed and water or pasture grass. Young ones hatch out in the intestine and penetrate the wall. These entere the blood steam and move up to liver and lungs. From the lungs they move upto mouth and swallowed again to be returned to intestine slowly where they mature in about 2 months.

Damage : Competing with host for nutrients, digestive, disorder, lung infestation causes difficulty in breathing, increases succeptibility to pneumonia in winter, lowers vitality, weakness.

Prevention : Disinfection of premises, change of pasture, clean and dry surroundings and houses, administration of piperazine, phenovis etc.

(ii) Lung Worms

Cause : About 2 cm long, white, slender and thread–like worms found in wind pipe.

Infestation : Large number of eggs are coughed up and swallowed. Then pass out with droppings. The lung worms are swallowed by angle worms in which they hatch and develop. Pigs eat angle worms with feed and get infested which penetrate intestine to enter the blood stream whereby they reach to heart and lungs to follow the cycle again.

Damage : Weakness, coughing, difficulty in breathing, severity, causes death.

Prevention : Rotation of patures, sanitary practices, isolation of sick animals feeding fortified well nutritious ration.

Worm Controls in Pigs :

Good basic hygiene, use of anthelmintics in feed like–benzimidazoles, levamisole, and dichlorvos as per instructions. Anthelminthic programme in pigs includes treatment of sows and gilts ~ 10 days before breeding and again before farrowing. Treatment of weaners before entering clean pens; treat boars at 6 months intervals. Alternatively, an injection of ivermectin, also effective against lice and mange mites.

6.3.15. Mange

Cause : Mites. It feeds on tissues of the skin and blood. It burrows into skin causing dry, rough and scaly hide.

Symptoms : Intense itching, rubbing of body against walls and fixtures, biting by hogs, body with red rashes and scaly hide.

Prevention : Proper cleaning of shelters, proper disposal of manure, disinfection of all surfaces with chemicals. Spraying of insecticides in the houses.

Treatment : Application of sulpher ointments.

6.3.16. Diarrhoea

This is common and sometimes fatal with young pigs. The causes are over–feeding the sow after farrowing and feeding of decomposed sour feeds. Prevention consists of change in feed, control diet just after farrowing, supply of lime water with 15 to 20 grains iron sulphate. Diarrhoea in new born pigs due to enteropathogenic strains of **Escherichia coli**. These isolates of *E. coli* were haemolytic and resistant of Panicillin and Tetra cyclines but sensitive of Furazolidone, Neomycin and Streptomycin.

Constipation : The causes are lack of succulent feeds, lack of bulky feeds, to much concentrates, insufficient exercise. Cure consists of drenching of 100 gm epsom salt. For prevention laxative feeds such as wheat bran should be included in the concentrate mixture.

6.3.17. Crippling

In this there is stiffness and lameness usually on the hind legs. The animal lies most of the time and walking becomes difficult or impossible. The causes are strong fibrous food, too much of rich food, lack of exercise, damp quarters due to poor ventilation, wet floors, and filth.

For curing give 120 gm of epsom salt in 500 ml water. Prevention consists of feeding succulent green laxative feeds.

6.3.18. Rheumatism

The symptoms of rheumatism are lameness, stiffness, pain and swelling in the joints. It is caused by damp quarters, wet floors, poor ventilation.

Treatment consists of giving 20 to 30 grains one dose of salicylate of soda three times daily in the feed. Liniments can be applied to the affected joints.

6.3.19. Iodine Deficiency

Iodine deficiency occurs due to lack of iodine in the feed or water supplied to pregnant sow.

Affected pigs are usually weak at birth and die within a few hours. Some may be born dead. Such pigs are usually hairless. There may be swelling due to enlargement of the thyroid gland in the region of the throat. Iodine deficiency may be prevented by adding 1 to 2 grains of potassium iodide to the daily ration of pregnant sow.

6.3.20. Antinomycosis in Pigs

Causes : Actinomycosis suis.

Symptoms : Primary chronic granulomatus and suppurative mastitis. Small abscesses on the udder contain, viscid, yellowish pus surrounded by a wide zone of dence connective tissue.

Treatment and control : not effective ; Infected mammary gland must be excised to save the life to sow and to make her suitable for slaught.

6.3.21. Agalactia Syndrome

It is failure of let down or production of milk occurs in individual sow. Sow may have hypogalactia or Agalactia in one or more glands.

Cause :

Teat necrosis, bacterial mastitis, blind teats, Klebsiella or staphylococci infection.

Symptoms : hard udders, lactational insufficiency.

Control : Prompt diagnosis and therepy as per vet advice. Cross fostering is essential to save piglets.

6.3.22 Pediculosis in Pigs

(house infestation/Lousiness).

Cause : *Haematopinus asini* (Hog louse)

Symptoms : Hog louse is found around ears, backs and sides of pigs. Lice puncture the skin and suck blood. Pigs scratch and rub themselves; Show irritation.

Treatmetn and Control : Regular inspection, application of insecticides like lindane and Methaxychlor preparations or other insecticides governed by local regulations. May be treated with 0.06% Coumaphos, 0.15% dioxathion, 0.05% malathion, throxychlor, permethrin etc.

6.3.23. Rotaviral Enteritis

It is common disease of the small intestine of pigs. All ages are susceptible but usually occurs in nursing and post weaning pigs.

Symptoms : Rotavirus destroys villous enterecytes in small intestine, lesions are severe in middle third of intestine; causes partial villous atropy, mal–absorption and osmatic diarrhoea in pigs 5 days to 3 weeks old yellow grey and pasty. Weaned pigs have watery faeces; emaciation.

Cause and transmission

Rota–virus, by direct contact, Helathy carrier sows may be fecal shedders during periparturient period, thereby exposure of litters to infection.

Treatment and control :

No specific Treatment ; Provide adequate water to prevent dehydration; Minimize heat loss; Vaccination of sows helpful. Strict hygiene measures.

Note : Concurrent infection by *E. coli* is common, therefore antibiotic therapy will reduce mortality.

6.3.24. Anthrax

Cause :

Bacillus anthracis ; a gram–positive non motile spore forming bacterium; Bacilli from on opened up carcass exposed to air from spores which are resistant to extremes of temperature, chemical disinfections and desiccation; can survive in soil for years.

Symptoms : Affects all warm blooded animals. Some pigs may die to acute anthrax without any sign. Others show high fever, inflammation of the throat, bloodstained fluid coming out from all body opening, severe respiratory distress and dysentry.

Prevention and control :

Anthrax is usually fatal; Preventive programme is essential; animal died of anthrax should be burned. Anthrax spore vaccine gives protection to healthy animals.

Note : Animals respond well to penicillin, if treated in early stages of disease. Oxytetracycline, given daily to also effective.

6.3.25. Pasteurellosis

Cause :

Pasteurella multocida

Sudden change in the environment causing stress to animal body makes prone to infection of this disease.

Symptoms : Depression, huddling of piglets, breathing quickly, discharge from nose and cough frequently, Reduced feed intake, general unthrifliness and poor growth.

Treatment and Control :

Avoid stress of providing proper environment, segregation, isolation of new stock before induction in herd. Sound management, good sanitation, good nutrition and provide proper ventilation in pens.

6.3.26. Leptospirosis

Cause : *Leptospira icterohaemorrhagiae* : The infection is usually spread by the urine of an infected animal. Other species are **pomona**, **gripportyphosa**, bratisalava, and **canicola**.

Symptoms : Fever and anorexia pasting 3 to 4 days in acute cases. Severe clinical symptoms are poor weight gain, anorexia, G.I. disturbances, spasms, rigidity and circling. Abortion in sows in 2 to 4 weeks farrowing, stillbirth, or weak neonates.

Prevention and Control

Avoid use of wet, marshy pasture area, vaccination of all breeding animals in herd, seek vet. help for treatment.

6.3.27. Swine Dysentry

Cause :

Treponema hyodysentriae is anaerobic sphirochete which produces a haemolysin. It affects the large intestine of pigs. Transmission through feed or water.

Symptoms : Anorexia, with/without fever, passage of soft faces; course is variable some pigs may die per acutely; most common have mucoid diarrhoea, with flecks of blood and mucus, then watery mucohemorrhagic diarrhoea; dehydration, profound weakness gaunt and emaciation.

Treatment and Control :

Water medication to prevent dehydration; Therapeutic use of antibacterial drugs like Bacitracin, carbadox, lincomycin, virginiamycin, tiamulin, nitromidazoles etc; Sanitation and hygiene; disinfection of premises.

6.3.28. Tuberculosis

Cause

Mycobacterium tuberculosis : Pigs are susceptible to all three types of tubercule bacilli *i.e.* **bovis, avium** and **tubeculosis**.

Symptoms : Disease affects all parts of body. Primary lesions are in G.I. tract and associated lymph nodes; Intermillent high fever. Pigs become very docile, sluggish and have good appetite.

Prevention and Control : Remove the source of infection, Periodic testing and removal of reacting animals; Sanitation and disinfection; vaccination of healthy animals.

6.3.29. Protozoal Infection in Pigs

Cause :

Eimeria spp. and *Isospora suis.* through contaminated feed and water.

Symptoms : Diarrhoea, dysentry, abdominal colic, vomiting, loss of appetite, muscular weakness, faeces with blood and mucus.

Treatment and Control : Treatment by use of Amprulium, Bifuran in feed, diguinol etc. Sanitary disposal of infected faecal material, Prevent contamination of feed and water.

6.3.30. Some Common Ailments in Hogs.

1. Salt poisoning :

Causes : Excess intake of salt through feed.

Symptoms : Ichiness, thirst and constipation; causing deafness, indifferent around the pen bumping into walls.

Prevention : Provide water at all times, replace fed, provide salt free ration for about 8 to 10 days.

2. Mercury poisoning :

Causes : Feeding grains treated with fungicides having mercury compounds.

Signs : Off the feed, suffer partial blindness, eventually become uncenscious and sudden death in severe poisoning.

Prevention : All surplus fungicides treated grain to burned and ashes burried deeply. Seek vet. help for treatment.

3. Lead poisoning

Causes : Pigs let loose to feed on garbage/waste heap with old paint cans, tins and batteries having residues of lead based compounds.

Signs : Off feed, pass blood stained stools, stagger, squeal, jaw munching, convulsions, partial blindness and coma–in final stages.

Prevention : First aid by providing 120 gm Epsom salt (magnesium Sulphate) in a solution form orally. Prevent easy access to the disposed cans, tins batteries on wastes.

4. Mold poisoning

Causes : Fusarium, Aspergillus, mucor poisonous variety.

5. Lameness and Paralysis

Causes : arthritis, lack of calcium and phosphorus, Vit. A and B deficiency, Poor management.

Signs : Swelling of joints, partial paralysis of individual limb or both rear legs, limping shifting of weight from leg to leg.

Prevention and Control : Providing balanced ration, ideal environment and good housing facilities.

6. Heat Stroke and Sunburns :

Causes : Excessive exposure to sunlight, extremely hot weather, poor ventilation, high humidity, lack of sufficient fresh water in summer, stranuous exercise.

Signs : High rise in body temperature, panting/gasping for breath, increase in respiration rate; In sunburns formation of pink area on body, reddened skin–may develop blisters, staggers and sways.

Control : Apply wet, cool pack to the back and affected parts of sides and head, provide ventilated housing, hfesh water at all times, outdoors pigs be provided shade and cool water. Prevent exposure to sun; application of soothing medicines as per vet. advice.

7. Mastitis

Causes : Insanitary conditions, bacterial/viral infection (streptococci, staphylococci, corynbacteria, *E. coli* etc.) Fungal infections, Injury to teats by piglets with needle teeth; *Fusobacterium necrophrum* etc.

Symptoms : Swelling of one or more quarters inflammation of udder, pain in affected teat, glands are hot and painful, stops milk flow, hardness on udder, fever loss of appetite, constipation.

Prevention and Control : Good hygiene and sanitation, removal of needle teeth of piglets to avoid injuries to udder, cleaning of teats and glands with antiseptic solution after suckling. Keeping floor dry. Treatment includes administration and broad spectrum antibiotics like ampicillin, dihydro–streptomycin and oxy–tetracycline.

8. Metritis

It is the inflammation of Uterus caused by bacterial infection due to insanitary conditions, Sows farrowing show signs of metritis.

Signs : Off–feed, depressed high temperature, reluctant to move, putrid discharge from genitals.

Control :Check spread of infection through natural breeding, sanitation and disinfection of pens. Rest for breeding, cleaning genitals with legal solution. Treatment as per vet. advice.

9. Haemolytic Disease of Newborn Piglets

Cause : Natural transplacental sansitization; Maternal sansitization after use of crystal violet–inactivated hog cholera vaccine.

Signs : 2 to 4 days after birth of piglets inactivity, dyspnea, jaundice and pallor. The signs subsides gradually. Then after 10 days piglets have petechial haemorrhages over ventral surface and die in most cases. RBC drops from normal.

Treatment and Control : Not effective; stop rebreeding sow with the same sire, use substitute dam for cross fostering.

6.3.31. Other common ailments and their possible causes

Ailment	Causes
1. *Abnormal skin and hair*	Salt and protein deficiency.
2. *Anaemia*	Deficiency of iron, copper and Vit. B.
3. *Birth of blind pigs*	Vit. A deficiency.
4. *Birth of dead or weak pigs*	Deficiency of Vit. A, iodine.

Ailment	Causes
5. *Birth of hariless piglets*	Deficiency of Iodine.
6. *Lameness of rickets, posterior paralysis in sows*	Lack of calcium, phosphorus and vitamin A, D & B
7. *Poor appetite*	Deficiency of protein, vitamins, iron and calcium.
8. *Poor reproduction (Infertility/low conception rate)*	Either overfatness or run down condition, deficiency of calcium, phosphorus, iodine and Vit A. & E deficiency.
9. *Scours, diarrhoea*	Deficiency of Vit. B, Poor sanitation, ration too fibrous.
10. *Slow growth*	Grain deficiency, too little feed, poor quality feed, minerals and vitamin deficiency; parasitic infestation, dysentry, respiration diseases, enteric disease.
11. *High mortality*	Transmissible gastroenteritis.

6.4 . METHODS OF DISEASE CONTROL

1. Segregation
2. Erection of physical barriers 50 meters around pig endosures to keep off visitors, cars and advisors equipments and surgeons, farm attendents etc.
3. Personal hygiene–
   Discouraging casual callers, visitors, use of protective clothings, disinfecting feet and hands.
4. Check on quality of feed and water, ensure feed is steam pelleted and is packed in paper sacks.
5. Limiting of pasture and keeping it parasite free.
6. Rotation of pastures.
7. Proper sanitation in house (dry floor, well ventilated and good lighting arrangement).
8. Proper disposal of infected litter and carcass.
9. Quarantine.
10. Vaccination of healthy animals.
11. Timely visit and treatments.
12. Periodical spray of insecticides in houses.

13. Providing balanced ration with low in fibre but with more succulent green leguminous laxative feed.
14. Provide facility of medicated wallow.
15. Keeping proper records for conception rates, farrowing index, number born, number weaned and reared, mortality, etc., to identify.

Vaccination Schedule of Swine

Disease and vaccine	*Age*	*Dose & Route*	*Booster Dose*	*Duration of immunity*
**Swine fever*	2 weeks	1 ml. I/M	Annual	One year
**Swine erysipelas*	10 weeks	1 ml. S/c	Annual	One year
**F.M.D.*	1 month	5-10 ml. S/c	21 days	6 months
**Brucellosis*	2 months	2 ml S/c	---	Life long
**H.S.*	3 months	2-3 ml S/c	Before rainy Season Annual	6 months
** Swine influenza*	At any stage	1 ml S/c	As per need.	6 weeks.

Disease diagnostic routine

1. Visit to the slaughter house to check proportion of carcasses.
2. Routine postmartum examination at regular intervals.
3. Examination of stock of regular periods.
4. Laboratory checks for salmonellosis, **E.coli**, brucellosis, leptospirosis and internal parasites.

Medication

Broad spectrum anthelmentics and parasiticides can be used routinely to reduce infection of both internal and external parasites to minimum.

Immunity

Commonly used vaccines in pigs are that of swine erysipalas and swine fever to save gaurd healthy pigs.

6.5 COLLECTION OF BLOOD FROM HOGS FOR HAEMOLYTICAL EXAMINATION

Animals		*Suitable vein for puncture*
1. Horse, cow, sheep, goat	–	Jugular vein
2. Cat, dog	–	Jugular vein and Saphenous
3. Rabbit, guinea pigs	–	Vein, Heart.
4. Pigs	–	Anterior Vena cava.

(a) Collection of blood from ear vein

1. Flip on ear vigorously with the hand and rub with xylol and alcohol.
2. Puncture a marginal vein with a large needle and collect the blood 2 to 3 ml as needed.
3. Check the bleeding by firm pressure over the puncture.

Note : Xylol produces marked congestion and afterwards should be removed with alcohol and water because it produces a low grade inflammatory reaction.

Sample from ear has not proved sufficiently reliable for accurate haematological studies.

(b) Blood collection by amputation of tail

1. Clean the tip of tail and wipe with alcohol.
2. Lay the end of tail on a block of wood and chop off about 1 cm with razor.
3. Allow the blood to run into sterile container.
4. When needed amount is obtained, tie a string around the tip of the tail to prevent further bleeding and seal wound with tincture benzoine.

Collection from vena cava

1. Restrain small pigs in the dorsal recumbent position. Restrain hogs in the standing position by means of a rope around the upper jaw that is snubbed to a post of stanchion.
2. Use a glass syringe and a 4 to 5 cm needle. A 20 gauge needle is recommended.
3. Insert the needle 2 to 4 cm (depending upon size of animal) from the apex of carinform cartilage on a line drawn from the point of the cartilage to the base of ear.
4. Guide the point of needle inward, downward, and backward (assuming the animal is dorsal recumbent) to the entrance of the thorax between the first pair of ribs.
5. The needle then pierces the vena cava at a point where the jugular and brachial vein converge. The blood flow readily into the syringe but with gentle suction.

5.6. MORTALITY OF PIGLETS

Survival rate of piglets between birth and weaning is an important factor which determines the economic viability or swine enterprise. Even in well managed farm 25 to 30 per cent of the piglets born never reach the weaning age. While the pre-weaning piglet mortality ranges from 20 to 30 per cent depending upon managemental condition. The post weaning mortality is

Pre weaning mortality of piglets

PER CENT

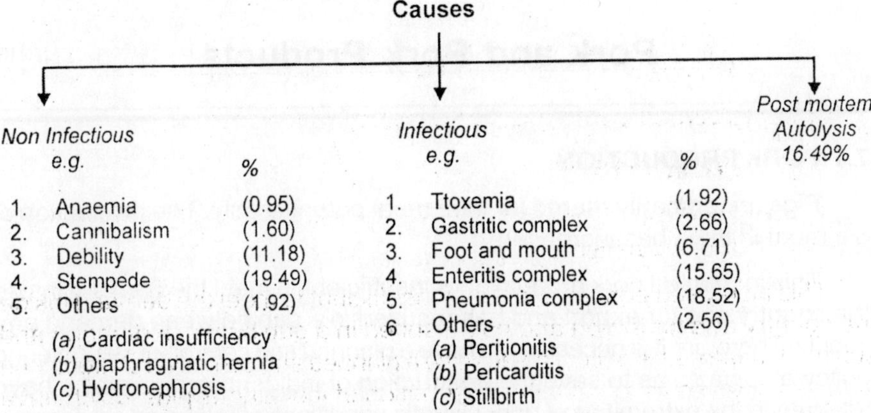

about 3 per cent.

Measures to control non–infectious causes of mortality

1. Improved management practices.
2. Good feeding.
3. Better management at breeding and gestation.
4. Provision of farrowing guards rail.
5. Constant vigil during first week of age.
6. Distribution of piglets among mothers.
7. Spraying ferrous sulphate saturated solution alongwith sugar on sows udder.
8. Free access of soil parasite free place.
9. Spraying insecticides periodically.
10. Better sanitation.

7

Pork and Pork Products

7.1 PORK PRODUCTION

Pigs are primarily reared for thier meat potential only. The production of pig meat in India has increased.

This increased pork production is insufficient to meet the demand inside the country and for export and has resulted in a gap between demand and supply. Therefore it is necessary to have a planned and controlled slaughtering policy of pigs so as to save the destruction of indigenous pigs which have adapted to the extremities of geo–climatic conditions of India and are famous for disease resistance and can be reared on poor quality feed. The export potentials of pig meat and its products have not been fully realized and efforts have not been made for export oriented production. There is a need to rear, slaughter and pack the pig and pig products which could satisfy international standards and norms of sanitation. This would help pig rearers realise better profits, which will in encourage their interest in pig industry.

By–products

The bristles, bones and fat are some of the important by–products that are used in preparation of brush, fertilizers, food products etc. Pig dung can also be used as manure.

7.2 SELECTION OF HOGS FOR SLAUGHTER

Parameter	*Desirable feature*	*Reasons for*
(a) Age	6-10 months	1. Pork of young pigs contains more water so it is nearly white and tender. It lacks firmness, flavour and keeping quality.

Parameter	*Desirable feature*	*Reasons for*
		2. Pork of old hogs is less tender and has rose colour. 3. Hogs of 6 to 10 months age produce pork of desirable tenderness firmness, colour of keeping qualities.
(b) Soundness	Hogs be free of injuries, wounds etc.	1. Meat of injured parts lack flavour and keeping quality. 2. Injured animals must be killed, bled and dressed or else blood will collect on injured parts and may cause swelling. 3. Meat of injured parts, if blood, gets collected, may become sour before or during curing process.
(c) Condition	Hogs should be medium of flesh.	1. Muscles of thin hogs are usually tough, lacks tenderness, juiciness and flavour. 2. Over fat/oily hogs do not produce good quality cured pork due to excess fat.
(d) Health	Free of TB, Hog cholera, parasites etc., irrespective of age quality of condition.	To prevent disease transmission to consumers. Healthy hogs produce meat of superior quality
(e) Type	Intermediate type	Old lard type hogs do not produce desirable kind of ham, shoulder and bacon.
(f) Quality	Smooth, evenly fleshed.	1. To get good quality meat and cut up with much less waste. 2. To get high percentage of edible meat, fine texture and tender quality meat. 3. Rough, coarse, wrinkled individuals produce poor quality meat with more waste.
(g) Sex	Castrated male and sow not near to farrowing stage.	To prevent strong, unpleasant odour and flavour of boar.
(h) Dressing Percentage	60 to 80 per cent	1. Hogs of 6 to 10 months age dress better and yield higher dressing percentage. 2. Dressing percentage depends on age condition, quality and style of dressing.

7.3 JUDGING AND GRADING OF LIVE PIG

Animal No........... Sex ....Age ....Breed...........
Weight ....

Scale of Points

I. General : 40 points

1. **Weight** : according to trade requirement 10
2. **Form** : Long compact, level back and neat shoulder 10
3. **Condition** : Well fleshed and suitable quality of fat 12
4. **Quality** : Fine skin, free from wrinkles, fine silky hair and fine in the bone. 08

II. Body Parts : 60 Points

1. **Quarters** : Long and wide 10
2. **Loin** : Broad and long 10
3. **Ribs** : Well sprung 08
4. **Flanks** : Deep and full 06
5. **Shoulders** : Neat and levelled 06
6. **Neck** : Well developed 04
7. **Head** : According to breed, wide light good 03
8. **Forelegs** : Straight and well placed 05
9. **Ham** : Wide and well fleshed to hocks. 08

Total 100

Recording to Observations

Sl. No. of Animals	General				Body Parts									Total
	1	2	3	4	1	2	3	4	5	6	7	8	9	
1														
2														
3														
4														
5														

7.4. TOOL REQUIRED FOR SLAUGHTER

1. Butcher knife with 20 cm long blade.
2. Bell shaped hog scraper.
3. Meat saw.
4. A hog gambrel.
5. A hog hock of 35 cm long.
6. Heating water trough/tank.
7. Scalding water.
8. Weighing balance.
9. Chopper.
10. A platform for scraping operations.
11. Physical weighting balance.
12. Polythene bags.
13. A long double edged knife.

7.5. STEPS FOR SLAUGHTER OF HOGS.

I. Management steps before butchering hogs.

1. Do not feed hogs 24 to 30 hours before killing but fresh clean water be offered free of choice for the following reasons.
   - *(a)* Prevents fermentation and formation of gases in stomach and intestine.
   - *(b)* Removes contents of stomach and intestine.
   - *(c)* Helps in cleaning of organs.
   - *(d)* Makes easier to get good bleed.
   - *(e)* Saves feed.
   - *(f)* Meat cures better because blood vessels are free of blood.
2. Animals before slaughter should not be excited or over heated because this may cause increase in body temperature which may cause pork to become sour.

II. Step for butchering

1. Pigs are made insensible by such means are as electric current, exe, gunshot, captive botl stunner/pistol.

2. **Shackling** : Hogs are secured from the hocks of hindlegs and then hoisted on overhead rails keeping its head down.

3. **Sticking** : Take a long double edged knife and insert just in front of the breast bones and directly over midline.

When knife is inserted 15 to 20 cm deep (according to size of the hog) it is given a quick turn and then withdrawn. This bleeds the hog to death. The animal is allowed to bleed for few minutes.

4. **Scalding** : Place the hog in scalding wate in a vat for few minutes at the temperature of 65°C. It helps to loosen the hair and scurf.

5. **Scraping** : After scalding hog is lifted on to a platform for removal of hair by scraping with candle stick stick scraper, or the pig is elevated on to a dehairing machine to remove hair mechanically. Remaining hair on legs or body may be scraped out but knife.

Note : Some people prefer to remove skin of the hogs rather than scalding and scraping.

6. **Hoisting and overhead rails** : The scraped out hog's carcass is returned to overhead rails and hanged by inserting the gambrel stick in the tendens on the back of the hind legs.

7. **Dressing** : It includes following steps :

(a) Washing and singering (burning bristles after killing).
(b) Removal of head.
(c) Opening the carcass and evisceration.
(d) Spliting the carcass with saw.
(e) Removal of fat.
(f) Exposing the kidney for inspection.
(g) Washing the carcass.
(h) Storing of pork at 1 to 2°C.

8. **Dressing percentage** : It depends upon age, sex, condition of hog, style and quality of dressing. It general the average dressing percentage is as follows :

Pigs	=	70 per cent	Goat	=	50 per cent
Cattle	=	55 per cent	Sheep	=	45 per cent.

7.6. CUTS FROM PORK CARCASS (SEE. FEG. 7.1)

The cuts are hind feet, ham locks, ham fatback, loin, spare ribs, belly or side, clear plate, Boston butt, shoulder, hocks, knuckles, fore feet and jowl.

7.7 SLAUGHTER WEIGHT

Singh et. al. (1986) reported average slaughter weights of castrated and uncastrated male Hampshire as 76.9 and 79.64 per cent respectively.

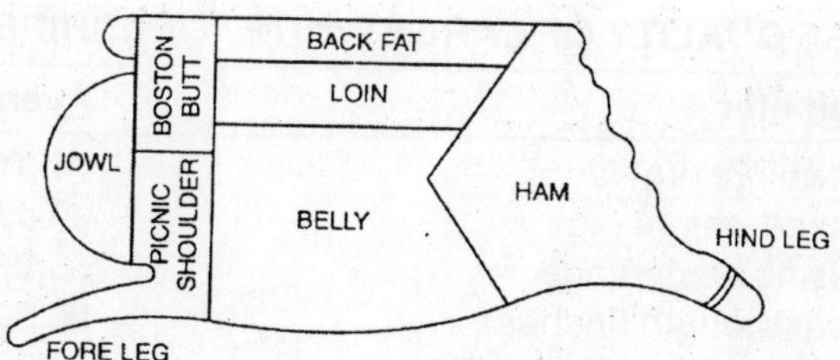

Fig. 7.1. Cuts of Pork

7.7.1 Slaughter Record of Pigs

Animal No............. Breed.............Sex ....Age.............date.............
Grade ....live weight (kg).........(Rs)........../kg...........Total (Rs.).............
Transport cost .... (Rs.)...
Cost of slaughter house (Rs.)..
Total ....
Live weight of slaughter (kg.)..........................Shrinkage (%) ....
Hot carcass weight (kg)..........................Cold carcass weight................

Offal	*Weight (kg)*	*Per cent of Live weight*	*Offal*	*Weight (kg)*	*Per cent of Live weight*
* Stomach + Content			Pluck		
			Melt		
* Stomach + (empty)			Caul fat		
* Intestine + content			Mesenteric fat		
			Miscellaneous		
* Intestine + (empty)			Miscellaneous		
* Trimmings etc.			Miscellaneous		
* Blood			Miscellaneous		
* Total			Total		

Cost of slaughter house (Rs.) ....
Add slaughter and storage cost (Rs.)...
Less value of offals..........................(Rs.)
Net cost of carcass (Rs.)...
Cost of carcass per kg.............(Rs.)
Signature.........................

7.8 CARCASS QUALITY OF LARGE WHITE YORKSHIRE PIGS

	Parameter	Average
1.	Live weight (kg)	74.31
2.	Dressed weight (kg)	49.47
3.	Dressing percentage	65.91
4.	Carcass length (inches)	26.59
5.	Back fat thickness (inches)	00.92
6.	Loin eye area (sq.cm.)	24.69
7.	Ham weight (kg)	10.25
8.	Loin weight (kg)	8.07
9.	Butt weight (kg)	4.93
10.	Picnic weight (kg)	5.63
11.	Per cent Ham in dressed carcass	20.36
12.	Per cent loin in dressed carcass	15.60
13.	Per cent Butt in dressed carcass	9.97
14.	Per cent Picnic in dressed carcass	11.43

7.9 JUDGING AND GRADING OF PIG CACASS

Animal No............Sex.............Age.............Breed............Weight..........

Scale of points

I. **General** : 40 points

1. **Weight** : According to trade requirements — 10
2. **Form** : long, levelled back, neat shoulders and short legs — 10
3. **Condition** : Well fleshed, suitable quantity of fat — 12
4. **Quality** : Small bone, fine skin, free from wrinkles kidney fat smooth, white and firm. — 08

II. **Carcass** : 60 Points

1. **Side** : long and deep, well fleshed and firm. — 15
2. **Shoulder** : Compact, not excessively fat. — 05
3. **Loin** : long, well fleshed and even thickness of fat throughout — 15
4. **Ham** : Short and plump, firm and correct covering of fat. — 12
5. **Belly** : thick and clean — 6
6. **Head** : light — 3
7. **Carcass** : Well dressed free from hair, scar and bruishes. — 4

Total 100

Recording of data

Sl. No. of Animals	General				Carcass Parameters							Total
	1	2	3	4	1	2	3	4	5	6	7	

7.10 PROPORTIONS OF EDIBLE AND INEDIBLE OFFALS

Prescott and Lamming (1964) and Vold (1969) reported that castration of pigs result in variation in proportions of various edible and inedible offals.

Edible Offals

1. **Heart** : It is recorded with pericardium.
2. **Liver** : It may be recorded with gall bladder.
3. **Spleen** : It must be separated from mesenteric fat and weighed.
4. **Kidneys** : These must be separated from renal attachments and surrounding fat and then weighed.

Inedible offals

1. **Stomach** : It must be separated from intestine at pyloric end, and then recorded.
2. **Intestines** : These must be weighed with their contents and mesenteric fat.
3. **Lungs** : These may be weighed with trachea.
4. **Head** : It may be weighed with the ears but wihtout jowl which must be removed during decapitation.
5. **Trotters** : It is the collective weight of fore and hind feets.

Means of per cent of edible and inedible offals : (Slaughter weight basis), Singh *et. al* (1986)

Various Offals	*Male Hampshire Pigs*	
	Castrated	*Uncastrated*
Slaughter Weight	76.9	76.64
1. **Edible Offals**		
(i) Liver	2.01	2.00
(ii) Heat	0.35	0.36
(iii) Kidneys	0.28	0.30
(iv) Spleen	0.16	0.17
Total	2.80	2.83
2. **Inedible offals**		
(i) Lungs	1.01	1.19
(ii) Stomach	2.30	2.37
(iii) Intestines	10.00	9.57
(iv) Head	5.85	6.08
(v) Trotters	2.29	2.42
Total	21.51	21.83

6.11 CURING THE PORK

Method	Amount of ingredients used for 100 kg meat	
1. Dry Salting	Common salt	– 8 to 10 kg
2. Dry cure mix	Common salt	– 8.0 kg
	Sugar	– 2.5 kg
	Pot. nitrate	– 70 gm
3. Sweet pickle cure (Brine method)	Common salt	– 8.0 kg
	Sugar	– 3.0 kg
	Pot. nitrate	– 80 g
	Water	– 27 litres

Procedure for curing pork

(i) **Dry method** : The first half of the mixture is rubbed thoroughly on pieces of pork and they are packed. Allow curing for 4 days. At the end of 5th day the remaining half of the mixture is also rubbed on the pieces and they are repacked. On 15th day the pieces are reworked.

(ii) **Brine method :**

(a) Brine is prepared with the ingredients mentioned earlier.
(b) Pack the pork skin side down into a clean hard wood barrel well soaked with water.

Note : Thicker and heavier cuts should be placed in the bottom of barrel.

(a) Sweet pickle cure/brine mix is poured over the pork until liquid just covers it.
(b) Some weight be placed over the pork to hold it under liquid.
(c) Allow curing for three days.

Note :

1. At the end meat is soaked in luke warm water for three hours and then hung for 24 hours in drip.
2. Curing room temperature should be 5 to 8°C.

7.12. SAUSAGE

The lean meat which is trimmed from the different parts generally goes into sausage. It con tains fat in proportions varying from 25 to 50 per cent of total weight. The meat is run through a chopper and mixed well before it is put in casings. The followings composition of mix is considered satisfactory under Indian conditions :

3 kg of meat
30 g salt
15 g black pepper
7 g nutmeg
7 g red pepper
14 g of sage

The sage and pepper may be increased or decreased to suit the market needs of individuals.

7.13. SMOKING OF CURED PORK

Objective : 1 Darkens the colour.
2. Increase the flavour.
3. Improves keeping quality.

Material : Hard wood saw dust.

Temperature for smoking : 57°C.

Procedure :

1. Soak the meat after curing in cold water for 2 hours.
2. Scrub the meat with the clean brush.
3. Hang the pork to dry over night in the smoking house to remove excess salt.
4. Hang the cuts apart from each other for three days.
5. Season it with black pepper according to taste or needs of the market.

Storage : It is not advisable to freeze the cured meat as it develops rancidity. When freezing is necessary storage period should not exceed more than two months.

Remark : Smoking house is specially constructed for the purpose and has a hold at the top of ventilation and a perforated base below for the fire. The fuel is allowed to burn slowly throughout for three days. Then all flames are subdued by sprinkling water on the fuel and cuts are taken out.

7.14 HEAD CHEESE

It is taken from the head but meat from legs, tongue and heart may also be added. The head is cleaned thoroughly by removing brain, eyes and ear drums.

Cook the meat in kettle at about 83°C until the meat separates from bones. Chop the meat into fine pieces and boil the contests and blend the mixture along with the liquid of meat (broth). Add the seasoning according to taste as the material is heated. Cook again for 15 to 20 minutes and pour into shallow pan and cool it. It is sliced and eaten–called head cheese which can be stored in a cool place for several weeks.

7.15 SCRAPPLE

it is flavoured dish prepared similar to head cheese along with corn meal, buck wheat or white flour. The meat is prepared as in case of head cheese and boiled alongwith broth. Add enough corn meal to make a mixture resembling thick mush. While adding corn meal keep on stirring to avoid lumps. Add also margoran and spices as per taste. Boil the mixture for 30 minutes and stir frequently to prevent sticking. Pour the hot scrapple into shallow pan ahd chill. Then slice it and fry. The slices may be dipped in egg solution before frying

7.16 CANNED PORK :

The pork to be canned is removed to bones and fat and cut into small pieces so that it fits well in cans. A steam presser cooker/autoclave/canner and a can sealer is essential for canning.

8

Economics of Pig Farming

8.1 OBJECTIVES

1. To get economically efficient production.
2. To get maximum returns.
3. To work out expenses and income accounts.

Inputs

(a) Finance, productive and healthy animals, house and farming materials.
(b) Feed water, medicines, etc.
(c) Labour

Outputs

(a) Live and slaughtered animals including byproducts.
(b) Manure.

Line of recommendation

1. Reduction of recurring expenditure.
2. Get maximum profit.
3. Recycle feed and finance.
4. Regular marketing.
5. Utilize locally available materials.

8.2 FACTORS AFFECTING ECONOMY OF PIG FARMING

1. Selection of site.
2. Supply of feed, water, etc.
3. Right choice of bread.
4. Expenditure on feed, labour, etc.
5. Principle of marketing.
6. Diseases and mortality.
7. Live birth and growth rate.

8.3 EXAMPLE

Economics of a Pig Farm with a Herd of 30 Sows

First Year

	Rs.
1. Input :	
A. Initial expenditure/fixed expenses	2,00,000
(a) Building	
(b) Animals :	
30 sows @ Rs. 3000 each	90,000
3 Boars @ Rs. 4000 each	12,000
(c) Incidental charges	1,000
Fixed cost –Total	3,03,000
B. Recurring expenditure :	
(a) **Feed**	**Rs.**
(i) Feeding cost for 33 animals for 12 months @ 2 kg per day @ Rs. 6 per kg (33 x 365 x 2 x 6)	1,44,540
(ii) 250 piglets up to 6 months age @ 1 kg/day @ Rs. 6 per kg. (256 x 18,500 x 1 x 6)	2,77,500
(iii) Green grass and recycling feeds (table wastes, agric, wastes etc., available on the farm)	Free
(b) Labourers 2 @ Rs. 1,200 per month (2 x 12 x 1200)	28,800
(c) Vaccines medicines etc.	3,000
(d) Miscellaneous (water electricity etc.)	5,000
Total recurring (a + b + c + d)	4,58,540
Grand Total (A + B) = 7,61,840	
II. Output : (During First Year)	
(a) Sale of 220 pigs at 6 months age (allowing mortality and infertility @ 10%) @ Rs. 50 per kg. (70 x 220 x 50)	7,70,000
(b) Sale of 50 tons manure@Rs. 200/ton	10,000
Total returns	7,80,000
Deduct expenditure of recurring=7,80,000–4,58,840 Gross income in first year	3,21,160
Allow Rs. 200 per day as farmers remuneration *i.e.* @ Rs. 6,000 per month	(–) 72,000
	2,49,160
Deduct Electricity, establishment cost & other expenses like water	4,160
	2,45,000
(–) Deduct bank Interest etc. @ 12% on fixed cost	36,360
Net Profit	Rs. 2,08,640

Recycle the first year's net profit for second year maintenance.

Second Year

	Rs.
I. Input	
A. Maintenance of Building	5,000
B. Recurring expenditure	1,44,540
(*a*) Feed for 33 animals	
(*b*) Feed for 33 piglets (farrow twice per year @ Rs. 6/ kg. per animal) To be sold every 6 months *i.e.* 600 piglets per year (600 x 183 days x 1 x 6)	6,58,000
(c) Labourers	28,800
(d) Vaccines and medicines etc.	5,000
(e) Equipment and miscellaneous	5,000
	8,46,340

If the first year's profit is reutilized during the second year, the above input will be proportionately reduced.

	Rs.
C. (*a*) Depreciation of buildings etc. 10%	20,000
(*b*) Interest 12%	36,360
(*c*) Hire of land and other expenditure	50,000
	1,06,360

Total expenditure during 2nd year = Rs. 9,52,700 (A + B + C)

II. Output 2nd Year	
A. 600 pigs (to be sold) every six months therefore at any time there will be 300 pigs only (600 x 70 kg x Rs. 50)	21,00,000
B. Manure 150 tones @ Rs. 200/ton	30,000
Total returns	21,30,000
Deduct expenditure	9,52,700
Gross income in 2nd year	11,77,300
(—) Allow farmers remuneration	(–) 72,000
	11,05,300
Detect other expenses	(–) 50,000
Income	10,55,300
Returns bank loan taken during Ist year for building etc.	(–) 3,03,000
Net Income	7,52,300
Seed stock worth (+)	1,50,000

Note : Part of the net profit may be utilised to purchase the land he hired and part will be recycled towards the next year's budget.

Seed stock worth Rs. 1,50,000 will still be available for another 3 years.

8.4 EXAMPLE

Economics of small unit of pig farming under S.F.D.A.

Programme for upliftment of weaker sections

For establishment of piggery unit of 3 to 4 sows and one boar subsidy in given @ 25 to 33% but not more than Rs. 6000 per beneficiary and 50% in case of scheduled caste/scheduled tribes, subject to maximum of Rs. 10,000 per beneficiary.

Financial Provision for 1987–88 in Uttar Pradesh

A budget provision of Rs. 116.12 lakh (including 50% share of Government in India) was available and sanction of the above amount was issued by the State Government for implementation of the programme during 1988–89.

The production programme in U.P. was launched in four districts of U.P. *i.e.* Meerut, Ghaziabad, Badaun and Partapgarh.

Assumptions

1. Period of schme	3 years
2. Age of gilt at the time of purchase ready for breeding	8 to 9 months
3. Cost per quintal of feed	Rs. 600
4. Number of litters to be produced per sow per annum	1.75
5. Average litter size	8
6. Average weaning size	6
7. Sale price of fattening pigs per kg. live Weight	Rs. 40
8. Subsidy by DRDA	
[District Rural Development Agency :	SF = 25%
M.F. = Marginal farmers	MF = 31.3%
Farmers and AL = Agric. Labourers]	AL = 31.3%
9. Rate of Interest	
(a) Bank Loan	25%
(b) Margin money (for SC and B.Cs)	6%

10. Laon from banks – The capital cost on housing and 3 sows + 1 boar to be recovered in 3 annual instalments. Rest of the loan is paid at the end of each year.

11. Insurance @ 4% on animals cost
    Cost of gilts and boar @ 3,000 each

ECONOMICS

A. Capital Cost

		Rs.
1.	Housing with cement flooring and asbestos/tile roof	5,000
2.	Cost of sows (3) + 1 boar @ Rs. 3000	12,000
	Total	17,000
	Subsidy 33.3@ by D.R.D.A.	5,661
	Money margin @ 20@ of SC's Corp. or B.Cs	3,400
	Total	9,061

Loan component on Capital cost of
Rs. 17,000 – 9,061 = Rs. 7,939

B. Operational Cost

1.	Cost of maintenance of 3 sws and 1 boar @ 1 kg. of pre-mixed/compound feed per day for one year @ Rs. 6 per kg feed (1 x 4 x 365 x 6)	8,760
2.	Supply of extra production @ ration per nursing sows and creep feed for piglets @ 210 kg per sow or 840 kg per unit @ Rs. 6 per kg feed per year 840 x 6	5,040
	(a) Transport charges on feed propulsion charges to be paid as bankers subsidy by – SFDA @ 33 1/3% on feed cost	4,600
	(b) Margin money @ 20% on feed cost	2,760
	(c) Loan component on operational cost = (8,760 + 5,040) + (4,600 + 2,760)	6,440

C. Flow of finance

1.	Total investment of unit 17,000 + 13,800	130,800
2.	Total Subsidy (5,661 + 4,600)	10,261
3.	Total margin money (3,400 + 2,760)	6,160
4.	Total loan to be advanced 7,939 + 6,440 = Rs. 14,339	

D. Receipts

1.	By sale of 18 pigs for slaughter (50 kg x 18 x Rs. 50/kg)	45,000
2.	Cost of pig manure (2kg/d x 12 adults + 365 days x @ Rs. 0.20)	1752
		46,752

Repayment schedule : Rs.

(a) Total loan on operational cost 6,440
(b) 1/3 of the laon component on capital cost 2,646
(c) Interest on total loan (interest for 1 year on loan for capital cost and for 6 months on for operational cost) (12@ on Rs. 7,939 + 1/2 (12% on 6,440) = (952+386) = 1,338

Total = 10,424

Profit at the end of first year
Rs. 46,752 – Rs. 10,424 = Rs. 36, 328

2nd year economics

1. Operations expenditure for 3 sows + boar and 2 batches of weaners to attain 50 kg live weight (18 + 18 + 3) = 8,760 x 5,040 x 2 =18,840
2. Insurance for 3 sows + 1 boar (out of 4%) of insurance permium 1% will be done by DRDA) 510

   19,350
3. Margin money @ 20% by SC/BC Corp 6,160
4. Loan component Rs. 19,350 + 6,160 13,190

Receipts in 2nd year :

1. Sale of 36 pigs for slaughter (3 x 50 x Rs. 50) 90,000
2. Sale of pig manure (2 kg/d x 21 Adults x 365 d x 0.2) 3,066

Total = 93.066

Repayment Schedule in 2nd year :

(a) Total loan in operational cost 13,190
(b) 2nd Instalment of loan on capital cost 2,646
(c) Interest on operational cost and left over capital 1,582 + 318 1,900

Total 17,736

Profit at the end of 2nd year
22,400 – 8,787 = Rs. 13,613
93,066 – 17,736 = Rs. 75,330

3rd year economics

1. Operational expenditure (3 + 1) and 2 batches of weaners to attain 50 kg live wt. (18 + 18 + 3 sows) 18,840
2. Insurance of 3 sows + 1 boar 510

   19,350

3. Margin money @ 20% by SC/BC Corp. 6,160
4. Loan component = 19,350 – 6,160 = 1,319

Receipts in 3rd year

(a)	Sale of 36 pigs (36 x 50 kg x Rs. 50)	90,000
(b)	Sale of manure	3,066
	Total	93,066

Repayment schedule in 3rd year

		Rs.
(a)	Loan on operational cost	13,190
(b)	Repayment of instalment of loan on capital	2,646
(c)	Interest on operational cost and left over capital 1,582 + 318	= 1,900
	Total	17,736

Profit at the end of 3rd year
93,066 – 17,736 = RS. 75,330

Abstract :

Profit During	Ist Year	=	36,328
	IInd Year	=	75,330
	IIIrd Year	=	75,330
	Rs.	=	1,86,988

Repayment of margin money

(a)	Total margin money released for 3 years 2,760 + 6,160 + 6,160	= 15,080
(b)	Total interest @ 12% for the margin money for 3 years (1,338 + 1,900 + 1,900)	= 5,138
	Total	20,218

Net profit = Total profit (—) Margin money and interest
= 1,86,988 – 20,218 = 1,66,770

Profit for 1 year = 55, 590

Profit per month = Rs. 4,638

8.5 EXAMPLE

Economics for Pig Unit Consisting 5 Sows and 1 Boar

A. Capital Expenditure

1. **Livestock cost**

Sow 5 @ Rs. 2,500	12,500
Boar 1 Rs. 3,000	3,000
	15,500

	Rs.
2. **Building :**	
Cost of constructions :	
Covered area (2 to 2½ m^2/pig = 30 sqft/adult pig)	
Uncovered area (2 m^1/2 pig) = 25 sq ft per adult pig	
Covered area = 30 x 6 = 180 sqft	21,600
@Rs. 120 per sqft = 180 x 120	
Uncovered area = 25 x 6 = 150 sqft	
@ Rs. 40/– sq.ft. = 150 x 40	6,000
Enclosure for 40 weaned pigs @ 10 sq/ft/pig	
40 x 10 = 400 sqft. @ Rs. 50/sqft	20,000
	47,600
	2,000
3. **Equipments :**	

B. Fixed Cost

Life of building = 50 years

Cost year $= \dfrac{47{,}600}{50} = 950$

Life of equipment = 5 year

Cost year $= \dfrac{2000}{5} = 400$

Interest on capital 47,600 + 2,000 + 15,500 = 65,100
@ 11% on 65,100 = 65,100 x 11/100 = 7,161
Total fixed cost = 950 + 400 + 7,161 = 8,511

C. Recurring expenditure :

1. Feeding Cost :

6 adults (6 months and above)				= 65.7 qt.
@ 6 kg/day = 18 kg x 365 = 6,570 kg				
40 young ones (in eadh of two farrowing)				
@ 1/2 kg/day = .5 x 40 x 45 days x 2 = 1,800 kg.				= 18 qt.
38 growers (2–3 months)				
38 x 1 kg. x 30 days	=	1,140 kg	=	11.4 qt.
38 x 1,25 x 30 days (3 – 4 months)	=	1,425 kg.	=	14.25 qt.
38 x 1.50 x 30 days (4 – 5 months)	=	1,710 kg.	=	17.10 qt.
38 x 2 kg x 38 days (5 – 6 months)	=	2,280 kg	=	22.8 qt.
				149.25 qt.
@ Rs. 6 per kg. = Rs. 600/qt. = 149.25 x 600				89,550

2. Labour Charges = 1,000 per month
   = 12 X 1,000 PM — 12,000
3. Electricity and water — 400
4. Medicines — 500
6. Miscellaneous — 500

Total recurring expenditure — 1,02,950

Fixed cost + Variable cost
1,02,950 + 8,511 = 1,11,461

Income

1. Sale of 40 piglets of 2 months age
   @ Rs. 600 each — 24,000
2. Sale of 8 growers of 78 kg. wt.
   @ Rs. 40 per kg.
   38 X 78 X 40 — 1,18,560
3. Value of 6 breeding stock
   @ Rs. 3,000 each — 18,000
4. Value of gunny bags
   230 bags @ Rs. 10 each — 2,300
5. Value of dung
   15 tons @ Rs. 200 per ton — 3000

1,65,860

Net profit in a year = Rs. 1,65,860 – 1,11,461 = 54,399
Per month profit = Rs. 4,528.0

8.6 EXAMPLE

ECONOMICS OF PIG FARMING FOR SUPPLY OF 500 KG PORK/WEEK

Expenditure/Fixed cost		Recurring Expenditure	
Particulars	*Amount Rs.*	*Particulars*	*Amounts Rs.*
A. **Capital :**		**A. Feed :**	
(i) Covered area = 3,170 sq ft.		(i) Boar (3)	
(ii) Uncovered = 8,640 sq ft.		4 X 3 X 365	= 4,380 kg.
11,810 sq ft.		(ii) Preg. sows (4)	
		3 X 4 X 365	= 4,380 kg.
10 % for road 1,181 sq ft.		(iii) Sows (25)	
12,991 sq ft.		2 X 25 X 365	= 18,250 kg.
@ 2,00,000/acre		(iv) Lact. Sows (4)	
= .2982 acre = Rs. 59,640		4 X 4 X 365	= 5,840 kg.
		(v) Young stock (130)	
		1 X 130 X 60	= 7,800 kg.

Expenditure/Fixed cost		Recurring Expenditure	
Particulars	*Amount Rs.*	*Particulars*	*Amounts Rs.*
B. Building :			
(i) Covered area 3,170 x @150 = Rs. 4,75,500		(vi) Fattening (120) 2 X 120 X 180	= 43,200
(ii) Uncovered area 864 X @ 30 = 25,920 Rs.		Total feed or 84 tons @ 6,000	= 83,850 kg. = 5,04,000
C. Equipment :			
(i) Grinder = Rs. 10,000		**B. Labour :**	
(ii) Feedcart @ 2,000 each (2 in No.)	= 4,000	3 Labour @ Rs. 1,000/p.m.	
(iii) Wheel Barrow	= 1,000	3 X 12 X 1,000	= 36,000
(@ 500 each 2 in No.)	= Rs. 2,000		
(iv) Slaughter Equipment	= Rs. 2,000	**C. Supply :**	
(v) Manure cart	= 2,000	@ Rs. 10/- animal/month	
(vi) Miscellaneous	= 1,000	10 X 286 x 12	= 34,320
	= 20,000		
D. Stock : E. Repair :		**E. Repair**	
33 Sows + 3 boars		Rs. 4/animal/month	
33 X @ 2,500	= 82,500	4 X 286 x 12	= 13,728
3 X @ 3,000	= 9,000	**F. Depreciation**	
	= 91,500	(i) Building — @ 10% on 4,75,500	= 47,550
		(ii) Uncovered area @ 10% on 25,920	= 2,592
Capital = 59,640 + 4,75,500 + 25,920 + 20,000 + 91,500 = 6,72,560		(iii) Equipment @ 10% on 20,000	= 2,000 = 2,000
		Total Rs.	52,142
		G. Interest :	
		On capital @ 10% on 6,72,560	= 67,256
		Recurring Total	= 7,07,446
	INCOME		
A. Sale of meat :	Rs.	Total Income	= 11,68,600
26,000 kg. of meat @ Rs. 40/kg.	= 10,40,000	Expenditure	(—) 7,07,446
B. Sale of replacers :		Net Income	4,61,154
70 X 20 X 40	= 56,000	Per month net income	= 4,61,154 ÷ 12 = Rs. 38,429.5
C. Sale of manure : 3 kg X 286 x 365	= 3,13,170 kg.		
or 313 tons @ 200 ton	= 62,600		
D. Sale of gunny bags :			
10 X 1,000	= 10,000		
Total	= 11,68,600		

9

Commercial Pig Farming

9.1. WHY DO PIG FARMING ?

The challenges faced by our country in securing the food as well as nutritional security to fast growing population need an integrated approach for livestock farming. Among the various livestocks species, piggery is most potential source of meat production and more efficient feed converters after the broiler. Apart from providing meat, it is also a source of bristles and manure. Pig farming will provide employment opportunities to seasonally employed rural farmers and supplementary income to improve their living standards. The advantages of the pig farming are :

a) The pig has got highest feed conversion efficiency *i.e.* they produce more live weight gain from a given weight of feed than any other class of meat producing animals except broilers.

b) The pig can utilise wide variety of feed stuffs *viz.* grains, forages, damaged feeds and garbage and convert their into valuable nutritious meat. However, feeding of damaged grains, garbage and other unbalanced rations may result in lower feed efficiency.

c) They are prolific with shorter generation interval. A sow can be bred as early as 8–9 months of age and can farrrowing twice in a year. They produce 6–12 piglets in each farrowing.

d) Pig farming requires small investment on buildings and equipments.

e) Pigs are known for their meat yield, which in terms in dressing percentage ranges from 65–80 in comparison to other livestock species whose dressing yields may not exceed 65%.

f) Pork is most nutritious with high fat and low water content and has get better energy value than that of other meat. It is rich in vitamins like thiamin, Niacin and riboflavin.

g) Pigs manure is widely used as fertilizer for agriculture farms and fish ponds.

h) Pigs store fat rapidly for which there is an increasing demand from poultry feed, soap, paints and other chemical industries.

i) Pig farming provides quick returns since the marketable weight of fatteners can be achieved with in a period of 6–8 months.

j) There is good demand from domestic as well as export market for pig products such as pork, bacon, ham, sausages, lard etc.

9. 2. SCOPE FOR PIG FARMING AND ITS CONTRIBUTION TO NATIONAL ECONOMY.

9. 2.1 The pig population of the country is 12.79 million as per the 1992 livestock census and constitutes around 1.30% of the total world's population. The state wise pig population are given in Annexure I. The pork production stands at 4.20 lakh tonnes (1995). During 1995–96 the value of pork and pork products was estimated to be Rs. 66584 lakh. The contribution of pork products in terms of value works out to 0.80% of total livestock products and 4.32% of the meat and meat products. The contribution of pigs to Indian exports is very poor. About 934 tonnes of pork and pork products were exported during 1995–96. The value of pork and pork products exported is Rs. 262 lakhs against the total value of Rs. 61604 lakhs on account of meat and meat products export.

9.2.2 The pig farming constitutes the livelihood of rural poor belonging to the lowest socio–economic strata and they have no means to undertake scientific pig farming the improved foundation stock, proper housing, feeding and management. Therefore, suitable schemes to popularise the scientific pig breeding cum rearing of meat producing animals with adequate financial provisions are necessary to modernise the Indian pig industry and to, improve the productivity of small sized rural pig farms.

9.2.3 In view of the importance of pig farming in terms of its contribution to rural poor and possible potentials for pig rearing in our country, Government of India has initiated measures to promote the pig farming on scientific lines under its five year plans. The first step towards this direction is establishment of eight bacon factories and organisation of pig production in rural areas attached to bacon factories. In order to make available good foundation stock, regional pig breeding stations were established for each bacon factory. Further expansion of pig breeding programmes paved the way for establishment of 115 pig breeding farms (1992–93) throughout the country. The location of bacon factories and pig breeding farms are given in Annexures II and III respectively.

9.3. FINANCIAL ASSISTANCE AVAILABLE FROM BANKS/NABARD FOR PIG FARMING

9.3.1 NABARD is an apex institution for all matters relating to policy, planning and operations in the field of agriculture credit. It serves as refinance agency for the ground level institutions/banks providing investment and production credit for various activities under agriculture and allied sectors and ensuring integrated rural development. It co–ordinates the development activities through a well organised Technical Services Department at the head office and Technical cells at each of the regional offices.

9.3.2 For undertaking the pig farming on scientific lines, loan from banks with refinance facility from NABARD is available. For obtaining bank loan the farmers/enterpreneurs should apply to the nearest branch of a Commercial, Co–operative or Regional Rural Bank in the prescribed application form, which is available in the branches of financing bank. Necessary help or guidance can be obtained from the technical officer attached to or the manager of the bank in preparing the project report, which is a prerequisite for sanction of the loan.

9.3.3 For piggery development schemes with very large outlays, detailed project report will have to be prepared. The items such as land development, construction of sheds and other civil structures, purchase of the breeding stock, equipment, feed cost upto the point of income generation are normally considered under bank loan. Other items of investment will be considered on need basis after providing the satisfactory information justifying the need for such items. The cost of land is not considered for loan. However, if land is purchased for setting up the piggery farm exclusively, it can be considered as beneficiaries margin money.

9.4. SCHEME FORMULATION

In case of commercial piggery units, the banks are expected to submit a project for availing the refinance. The scheme normally should include information on land, livestock markets, availability of water, feeds, vaterinary aid, breeding facilities, marketing aspects, training facilities, experience of the farmer and the type of assistance available from State Government Regional Pig breeding centres.

The scheme should also include information on the number of and types of animals to be purchased, their breeds, production performance, cost and other relevant input and output costs with their description. Based on this, the total cost of the project, margin money to be provided by the beneficiary, requirement of bank loan, estimated annual expenditure, income, profit and loss statement, repayment period, etc. can be worked out and included in the project cost.

9.5. REQUIREMENTS OF A GOOD PROJECT

A format prepared by NABARD for formulation of piggery development schemes is given in Annexure IV. The scheme so formulated should be submitted to the nearest branch of bank. The bank's officers can assist in preparation of the scheme or filling in the prescribed application form. The bank will then examine the scheme for its technical feasibility and economic viability.

A. *Technical Feasibility – This would briefly include :*
   - i) Nearness of the selected area to financing bank's branch.
   - ii) Availability of good quality animals in nearby livestock markets/ breeding farms.
   - iii) Source and availability of training facilities.
   - iv) Availability of concentrate feeds and kitchen/hotel/vegetable market waste and Food Corporation Godowns.
   - v) Availability of medicines, vaccines and veterinary services etc.
   - vi) Availability of veterinary aid/breeding centres and marketing facilities near the scheme area.
   - vii) Reasonability of various production and reproduction parameters.

B) *Economic Viability – This would briefly include :*
   - i) Unit Cost – The average cost of piggery breeding stock for some of the States is given in Annexure V.
   - ii) Input cost for feeds, veterinary aid, insurance labour charges etc.
   - iii) Output costs *i.e.* sale price of fatteners, piglets and culled animals.
   - iv) Income-expenditure statement and annual gross surplus.
   - v) Cash flow analysis.

C. *Bankability*

Repayment schedule (*i.e.* repayment of principal loan amount and interest).

Other documents such as loan application forms, security aspects, margin money requirements etc. are also examined. A field visit to the scheme area is undertaken for conducting a techno–economic feasibility study for appraisal for the scheme. The economics of piggery unit is given is Annexure VIa–VIf.

9.6. SANCTION OF BANK LOAN AND ITS DISBURSEMENT

After ensuring technical feasibility and economic viability, the scheme is sanctioned by the bank. The loan is disbursed in stages against creation of specific assets such as construction of sheds, purchase of equipments and animals. The end use of the fund is verified and constant follow-up is done by the bank.

9.7. LENDING TERMS – GERMS

9.7.1 Unit Cost

Each regional office (R.O.) of NABARD has constituted a State Level Unit Cost Committee under the chairmanship of RO–in–charge and with the members from development agencies, commercial banks and cooperative banks to review the unit cost of various investments once in six months. The same is circulated among the banks for their guidance.

9.7.2 Margin Money

NABARD has defined farmers into three different categories and where subsidy is not available, the minimum down payment as shown below is collected from the beneficiaries.

Sr. No.	*Category of Farmer*	*Farmer level of predevelopment return to resources*	*Beneficiary's Contribution*
a)	Small farmers	Utp Rs. 11,000	5%
b)	Medium farmers	Rs. 11,001 to Rs. 19,250	10%
c)	Large farmers	Above Rs. 19,251	15%

9.7.3 Interest Rate

As per the RBI guidelines the present rates of interest to the utilimate beneficiary financed by various agencies are as under :

Sr. No.	*Loan amount*	*CB's & RRB's*	*SLDB/SCB*
a)	Upto and inclusive of Rs. 25,000	12.0%	As determined by SCB/SLDB subject to Min. of 12% (for all laon slabs)
b)	Over Rs. 25,000 and upto Rs. 2 lakhs.	13.5%	–do–
c)	Over Rs. 2.0 lakhs	As determined by CB/RRB	–do–

9.7.4 Security

Security will be as per NABARD/RBI guidelines issued from time to time.

9.7.5 Repayment Period of Loan

Repayment period depends upon the gross surplus in the scheme. The loans will be repaid in suitable half yearly/annual instalments usually within a period of about 5–6 years with a grace period of one year.

9.7.6 Insurance

The animals may be insured annually or on long term master policy, where ever it is applicable. The present premium rate for non IRDP schemes is 6% per annum.

9.8. PACKAGE OF MANAGEMENT PRACTICES RECOMMENDED FOR COMMERCIAL PIG FARMING

Modern and well established scientific priciples, practices and skills should be used to obtain maximum economic benefits from pig farming. Some of the major norms and recommended pracctices are given hereunder:

I. Housing Management

1. Construct shed on dry and properly raised ground.
2. Avoid water–logging, marshy and heavy rainfall areas.
3. The side walls of the sheds should be 4–5 ft. high and remaining height should be fitted with GI pipes or wooden poles.

4. the walls should be plastered to make them damp proof.
5. The roof should be atleast 8–10 ft. high.
6. The pig stys should be well ventilated.
7. The floor should be pucca/hard, even, non–slippery, impervious, well sloped (3 cm per metre) and properly drained to remain dry and clean.
8. A feed trough space of 6–12 inches per pig should be provided.
9. The corners of feed troughs, drains and walls should be rounded for easy cleaning.
10. Provide adequate open space for each animal *i.e.* double the covered area.
11. Provide proper shade and cool drinking water is summer.
12. Dispose of dung and urine properly.
13. Individual pens for baors/lactating sows should be constructed.
14. The dry sows/fatteners can be housed in group pens.
15. Give adequate space for the animals. (The housing space requirement of pigs in various categories/age groups is given in Annexure VII).

II. Selection of Breeding Stock :

1. Immediately after release of the loan, purhcase the stock from a reliable breeder or from nearest livestock market.
2. For commercial pig farming upgraded/cross bred or exotic stock in good health should be selected.
3. While selecting a gilt or sow primary aim should be secure a female that will produce large suvivable litter and which can attain marketable weight at an age of six months or less. This can be done with the help of pedigree records/Veterinarian/Bank's Technical Officer.
4. Purchase animals which are ready to be bred.
5. Identify the newly purchased animal by giving suitable identification mark (ear notching or tattooing).
6. Vaccinate the newly purchased animal against diseases.
7. Keep the newly purchased animal under observations for a period of about two weeks and then mix with the other animals.
8. Purchase a minimum economical unit as suggested.
9. Purhcase animals in two batches at the interval of three months.
10. Follow judicious culling and replacement of animals in a herd.
11. Cull the old animals after 10–12 farrowing.

III. Feeding Management

1. Feed the animals with best feeds.
2. Give adequate concentrates in the ration.
3. Provide adequate vitamins and minerals.
4. Provide adequate clean water.
5. Give adequate exercise to the animals.
6. The feeding of the piglets is more critical and high quality and more fortified diets are needed for feeding them.
7. Feeding of the sows during pregnancy is utmost important for increased litter size.
8. The feed requirements of lactating sow varies with the size of the litter, weight, size and age of sow.
9. Commercial pig farming should aim at the exploitation of nonconventional feed resources *viz.*, waste from Kitchen/hotel/cold storage/warehouses, in replacing the balanced rations to minimise the cost of production.
10. The feeding regime adopted should take care of all the nutrient requirements of various categories of pigs. The nutrient requirements of breeding stock and growing pigs are given in Annexure VIII a and VIII b respectively.

IV. Protection Against Diseases :

1. Be on the alert of signs of illness such as reduced feed intake, fever, abnormal discharge or unusual behaviour.
2. Consult the nearest vaterinary aid centre for help if illness is suspected.
3. Protect the animals against common diseases.
4. In case of outbreak of contagious diseases, immediately segregate the sick and the healthy animals and take necessary disease control measures.
5. Deworm the animals regularly.
6. Examine the faeces of adult animals to detect eggs of internal parasites and treat the animals with suitable drugs.
7. Wash the animals from time to time to promote sanitation.
8. Strictly follow the recommended vaccine schedule as given in Annexure IX.

V. Breeding Care :

1. Pigs are highly prolific in nature and two farrowing in a year should be planned by adopting optimal management conditions.

2. For every 10 sows one boar must be maintained for maximum fertility.
3. Breed the animals when it is in peak heat period (*i.e.* 12 to 24 hours of heat).

VI. Care during Pregnancy :

Give special attention to pregnant sows one week before farrowing by providing adequate space, feed, water etc. The sows as well as farrowing pens should be disinfected 3–4 days before the expected date of farrowing and the sows should be placed in the farrowing pen after bedding it properly.

VII. Care of Piglets :

1. Take care of new born piglets by providing guard rails.
2. Treat/disinfect the navel cord with tincture of iodine as soon as it is cut with a sharp knife.
3. Feed on mothers milk for first 6–8 weeks along with creep feed.
4. Protect the piglets against extreme weather conditions, particularly during the first two months.
5. Needle teeth should be clipped shortly after birth.
6. Vaccinate the piglets as per recommended vaccination schedule.
7. Supplementation of iron to prevent piglet anaemia is necessary.
8. The piglets meant for sale as breeder stock must be reared properly.
9. Male piglets not selected for breeding should be castrated preferably at the age of 3–4 weeks which will prevent the boar odour in the cooked meat thus it enables production of quality meat.
10. Additional feed requirements of lactating sow must be ensured for proper nursing for all the piglets born.

VII. Marketing :

The marketable products of the piggery farming includes in piglets as breeding stock, piglets as fatterners, marketable fatteners and culls. The marketing avenues for the above products are like satelite fattening farms/breeding-cum-rearing farms and pork consumption centres. In order to promote the consumption of pork it should be supplied to the consumers in an attractive form. Therefore availability of either slaughtering facilities or bacon factories are to be ensured to convert the fatteners into wholesome pork and their products. The bacon factories that are being operated in our country are furnished in the Annexure II. The sale of piglets at 2–3 months of age will yield quick returns and enables the pig farmer to concentrate their efforts on maximizing the productivity of breeder stock. The other marketing strategy can be rearing of piglets upto marketing age for their sale as fatteners. Based on the market demand appropriate marketing strategy must be adopted in consultation with the local animal husbandry department officials.

ANNEXURE – I

STATEWISE PIG POPULATION IN INDIA (1992)

(in thousands)

Sl No.	*States/U.Ts.*	*Total*
	STATES	
1.	Andhra Pradesh	648
2.	Arunachal Pradesh	239
3.	Assam	1364
4.	Bihar	1125
5.	Goa	90
6.	Gujarat	103
7.	Haryana	517
8.	Himachal Pradesh	7
9.	Jammu & Kashmir	12
10.	Karnataka	380
11.	Kerala	135
12.	Madhya Pradesh	729
13.	Maharashtra	375
14.	Manipur	383
15.	Meghalaya	294
16.	Mizoram	112
17.	Nagaland	526
18.	Orissa	585
19.	Punjab	100

20.	Rajasthan	253
21.	Sikkim	45
22.	Tamil Nadu	672
23.	Tripura	188
24.	Uttar Pradesh	2905
25.	West Bengal	954
Union Territories		
1.	A & N Islands	36
2.	Chandigarh	4
3.	Dadra & N. Haveli	–
4.	Deman & Diu	–
5.	Delhi	12
6.	Lakshadweep	–
7.	Pondhichery	–
	ALL INDIA	12794

Note '—' indicates less than thousand.

ANNEXURE – II

LIST OF BACON FACTORIES

S. No.	*State*	*Capacity No. of pigs/day)*	*Address*
1.	Uttar Pradesh	100	Bacon Factory Central Dairy Farm, Aligarh
2.	West Bengal	20	Bacon Factory Harringhatta – 721 436 Mohanpur, Nadia
3.	Andhra Pradesh	100	Government Bacon Factory Gunnavaram – 521 101 Krishna Dist.
4.	Bihar	50	Government Bacon Factory Kanke – 834–005 Ranchi
5.	Maharashtra	100	MAFCO Bacon Factory. National Park, Borivili, Mumbai – 400 022
6.	Rajasthan	50	Meat Complex Alwar – 301 001
7.	Kerala	50	Meat Products of India, Koothatukulam – 686 662 Ernakulum
8.	Punjab	20	Pork Processing Plant, Kharar – 140 301

Annexure – III

STATEWISE –LOCATION OF PIG BREEDING FARMS

S. No.	*State*	*Location of breeding farms*
1.	Andhra Pradesh	Gannavaram, Gopannapalem, Muktalya, Padavagi, Tirupathi, Vishakapatnam
2.	Arunachal Pradesh	Karsingsa, Loiliang
3.	Assam	Diphu, Haflong, Kailiapani, Khanapara (University), Kanapara, (ACRIP), Khanapara (Govt.) Khanikar, Marigaon
4.	Bihar	Gaurikarma, Hotwar, Jamshedpur, Kanke
5.	Dadra & Nagar Haveli	Port Sailvasa
6.	Goa	Curti Ponda, Ela
7.	Haryana	Ambala City, Hissar
8.	Karnataka	Hassarghatta, Koila, Kudige
9.	Kerala	Ankamaly, Koothattukulam, Kunnamkulam, Mannuthi, Mundayal, Parasala, Thalayda Paramdu
10.	Madhya Pradesh	Baster, Jabalpur, Sakalo
11.	Manipur	None Tamenglong, Senepati North, Tarang, Torbumg
12.	Meghalaya	Baghmora, Dalu, Jowai, Mairang, Mawryngkneg, Nongstoin, Pynursla, Rongjeng, Rongkhon
13.	Mizoram	Kolasih, Lunglei, Selesih, Thenzawl
14.	Nagaland	Alukute, Medziphema, Merang, Phek, Suthazu, Tijit, Tunesand, Wokha
15.	Orissa	Bhaminagar, Chiplima

16.	Punjab	Badal, Chhaju Majra, Gurdaspur, Jalandar, Ludhiana, Maltowara, Malwal, Nabha
17.	Rajasthan	Alwar, Bharatpur
18.	Sikkim	Gyalsing, Tadong
19.	Tamil Nadu	Chettinad, Hosur, Pudukottai, Saidpet, Udagamandaam
20.	Tripura	Amarpur, Birchandramanu, Gandhi gram, Mendihihaor, Nabincherra, Nalkata
21.	Uttar Pradesh	Aligarh, Arzilins, Barabanki, Basti, Dehradun, Izzatnagar, Lalitpur, Moradabad, Nilgaon
22.	West Bengal	Bijanbari, Hainghatta, Pedong, Singruntaum, St. Mary's Hill Turki

Annexure – IV

FORMAT FOR SUBMISSION OF SCHEMES

Scheme : Commercial Pig Farming

1. GENERAL

i) Name of the sponsoring bank

ii) Address of the controlling Office sponsoring Scheme

iii) Nature and objectives of the proposed schemes

iv) Details of proposed investments

S. No.	*Investment*	*No. of units*
(a)		
(b)		
(c)		

v) Specification of the scheme area (Name of District & Block/s)

S. No.	*District*	*Blocks*

Annexure - IV (Contd.)

vi) Names of the financing bank's branches

S. No.	*Name of the Branch*	*Disctrict*
(a)		
(b)		
(c)		

vii) Status of beneficiary/ies Partnership/Company/ Corporation/Co-operative Society/Others)

viii) In case of area based schemes, coverage of borrowers in weaker sections (landless labourers, small, medium & large farmers as per NABARD's norms, SC/ST etc.)

ix) Details of barrowers profile (Not applicable to area based schemes)

(a)	Capability	
(b)	Experience	
(c)	Financial soundness	
(d)	Technical/Other special qualifications	
(e)	Technical/Managerial staff and adequacy thereof	

Annexure – IV (Contd.)

2. TECHNICAL ASPECTS

a) Animals

i)	Proposed Breed	
ii)	Age of the animal	
iii)	Arrangements for vaccination, identification and health certificate	
iv)	Insurance	
v)	Cost of Boar/Sows/Pigs	

b) Production parameters

i)	Age at first Farrowing	
ii)	Farrowing interval	
iii)	Farrowing percentage	
iv)	Number of piglets produced	
v)	Mortality of adults/young ones	
vi)	Age at which piglets/fatteners are sold	
vii)	Body weight of animals	

c) Herd projection – For big units only, (with all assumptions)

d) Housing

i)	Type of housing	
ii)	Floor space-adults/ piglets/fatteners	
iii)	Cost of construction	
iv)	Other civil structures (for commercial units)	

Annexure - IV (Contd.)

e) Equipment needed

i)	Water troughs	
ii)	Feeding troughs	
iii)	Other equipments	

f) Comments on technical feasibility.

g) Government restrictions, if any.

3. FINANCIAL ASPECTS

Sr. No.	*Name of Investment*	*Size of Unit*	*Unit cost with component wise break-up (Rs.)*	*Whether approved by state level unit cost committee*
			a)	
			b)	
			c)	
			Total	

ii) Down payment /margin subsidy (indicate source & extent of subsidy)

Annexure - IV (Cont.)

iii)

Year	*Invest-ment*	*No. of units (Rs.)*	*Unit cost (Rs.)*	*Total outlay (Rs.)*	*Margin (Rs.)*	*Bank loan (Rs.)*	*Refinance assistance (Rs.)*
1	2	3	4	5	6	7	8
Total							

iv) Financial vaibility (comment on the cash flow projection on a farm model/unit and enclose the same)

Particulars
a) Internel Rate of Return (IRR) :
b) Benefit Cost Ratio (BCR) :
c) Net Present Worth (NPW) :

v) Financial position of the borrowers (to be furnished in case of corporate bodies/partnership firms).

a) Profitability ratio

i) GP ratio

ii) NP ratio

b) Debt equity ratio

c) Whether Income tax & other tax obligations are paid upto date

Annexure – IV (Contd.)

d) Whether audit is upto date (enclose copies of audited financial statements for the last three years).

vi) Lending Terms :

i) Rate of interest.

ii) Grace period.

iii) Repayment period.

iv) Nature of Security.

v) Availability of Government guarantee wherever necessary.

4. INFRASTRUCTURAL FACILITIES

a) Availability of animals :

i) Source ii) Place of purhcase iii) Distance iv) Type of arrangements for purchase v) Availability in required numbers.	

Annexure - IV (Contd.)

b) Feeding :

i) Types of feeds ii) Source iii) Cost/animal/year	

c) Breeding/Veterinary services

i) Source ii) Place iii) Distance iv) Type of services available v) Availability of staff vi) Cost/animal/year	

d) Marekting :

i) Source for piglets fatteners and culled animals ii) Place iii) Distance iv) Price realised (Rs. per animal or Kg.) - Culls - Piglets - Fatteners - Pork etc.	

Annexure - IV (Cont.)

e) Other aspects

i) Source of technical guidance ii) Trainign facilities - Source - Periodically - Duration iii) Other Government support	

f) Supervision and Monitoring arrangement available with bank.

Annexure – V

STATEWISE AVERAGE UNIT COST OF BREEDING STOCK

(Rupees/Animal)

Sr. No.	*Name of the State*	*Breeding stock*			
		Boar		*Sow*	
1	*2*	*3*		*4*	
1.	Andhra Pradesh	1000		800	
2.	Bihar	1500	(cross bred)	1200	(cross bred)
3.	Gujarat	700		350	
4.	Haryana	375	(desi)	375	(desi)
		2000	(exotic)	1500	(exotic)
5.	Himachal Pradesh	1500		1000	
6.	Jammu and Kashmir	–		–	
7.	Karnataka	700	(3 months)	700	(3 months)
8.	Kerala	400	(2–3 months)	400	(2–3 months)
9.	Madhya Pradesh	1100		500	
10.	Maharashtra & Goa	1250		1000	
11.	***N.E. States***				
i)	Arunachal Pradesh	650	(2–3 months)	650	(2–3 months)
ii)	Assam	1000		500	
iii)	Manipur	750	(2–3 months)	750	(2–3 months)
iv)	Meghalaya	800	(2–3 months)	800	(2–3 months)
v)	Mizoram	1000	(2–3 months)	1000	(2–3 months)
vi)	Nagaland	800	(2–3 months)	800	(2–3 months)
vii)	Tripura	500	(2–3 months)	500	(2–3 months)
12.	Orissa	1000	(cross bred)	700	(cross bred)
13.	Punjab	375	(desi)	375	(desi)
		2000	(CB/exotic)	1500	(CB/exotic)
14.	Rajasthan	1500	(cross bred)	700	(desi)
15.	Tamilnadu	2500	(exotic)	2500	(exotic)
		300	(2–3 months)		
16.	Uttar	1300	(cross bred)	1100	(cross bred)
17.	West Bengal	2000	(cross bred)	1400	(desi)

Note : The above units costs are indicative in nature.

Annexure – VIa

ECONOMIC OF PIG FARMING – AT A GLANCE

1.	Unit Size	:	10 Sows with 1 Boar
2.	System of rearing	:	Semi intensive system
3.	State	:	Karnataka
4.	Unit Cost (Rs.)	:	136739
5.	Bank loan (Rs.	:	116200
6.	Margin money (Rs.)	:	20539
7.	Repayment period (years)	:	5 with one year grace period
8.	Interest rate (%)	:	13.5
9.	BCR at 15% DF	:	1.32 : 1
10.	NPW at 15% DF (Rs.)	:	106742
11.	IRR (%)	:	> 50

Annexure VIb
ECONOMICS OF PIG FARMING - INVESTMENT COST (10 SOWS + 1 BOAR)

Sr. No.	*Particulars*	*Specification*	*Physical Units*	*Unit Cost (Rs./unit)*	*Total Cost (Rs.)*
1	Sheds and other structures				
	a) Farrowing pens (4)	100 sft. per	400 sft.	50	20000
	lactating sow				
	b) Boar cum service pen	70 sft. per boar	70 sft.	50	3500
	c) Dry sow pens (6)	20 sft. per dry sow	120 sft.	50	6000
	d) Fattener shed -I	10 sft. per fattener	200 sft.	50	10000
	e) Fattener shed - II	15 sft. per fattener	300 sft.	50	15000
	f) Store room			50	9000
2	Water supply system (Bore well, electric motor pumpset – 1HP, water tank, pipeline)				
3	Cost of equipment	Lumpsum			14000
4	Cost of breeding stock :	Lumpsum			1500
	a) Cost of sows				
	b) Cost of boar		10	1400	14000
			1	1600	1600

Annexure – VIb (Contd.)

Sr. No.	*Particulars*	*Specification*	*Physical Units*	*Unit Cost (Rs./unit)*	*Total Cost (Rs.)*
5.	Capitalisation of recurring expenses for first one year				
	a) Breeder feed cost	3 kg per boar	12208 kg		
		3.5 kg per sow		0.5	4273
		70% kitchen garbage	8545.25 kg	4	14649
		30% conc. feed	3662.25 kg	5	5400
		0.2 kg per piglet/day	1080 kg		
	b) Piglet feed cost	1.5 kg per fattener/day	2700 kg		
	c) Ist batch of fattener feed cost	70% kitchen garbage	1890 kg	0.5	945
		30% conc. feed	810 kg	4	3240
	d) Insurance cost	6% of breeding stock cost			936
				900	10800
			1		468
	e) Labour wages		117 animal		480
	f) Cost of medicine etc.		240 months	4	
	For breeder stock			2	
	for weaners/fatteners				468
	g) Misc. expenses		117 animals	4	480
	for breeder stock		240 months	2	136739
	for weaner / fatteners				20511
6.	Total financial out lay (TFO)			Say	20359
					116228
7.	Margin money @ 15% of TFO			Say	116200
8.	Bank loan @ 95% of TFO				

Annexure – VIc

ECONOMICS OF PIG FARMING – TECHNO ECONOMICS PARAMETERS

1.	No. of sows (6–7 months old)	10
2.	No. of boars	1
3.	No. of batches	2
4.	Interval between two batches (months)	3
5.	No. of farrowings per year	2
6.	No. of piglets pe sow per farrowing	11
7.	Mortality among piglets	20%
8.	Mortality among fatteners	10%
9.	Mortality among adults is not considered as insurance cover is available	–
10.	Weaning period (months)	2
11.	Space requirement (s.ft.)	
	Boar	70
	Lactating sow with it's piglets	100
	Dry sow	
	Fattener of 3–5 months age	10
	Fattener of 6–8 months age	15
12.	Store room (s.ft.)	100
13.	Supplementary feed requirement (kg/day)	
	Boar	3
	Sow	3.5
	Weaner	0.2
	Fattener (3–5 months age)	1.5
	Fattener (6–8 months age)	2
14.	Concentrate feed % of total feed	30
15.	Kitchen garbage % of total feed	70
16.	Cost of construction of sheds (Rs./s.ft.)	50
17.	Cost of construction of store room (Rs./s.ft.)	90
18.	Cost of boar (Rs.)	1600

19.	Cost of sow (Rs.)	1400
20.	Cost of weaner feed (Rs./kg)	5
21.	Cost of concentrate feed (Rs./kg)	4
22.	Cost of kitchen garbage (Rs./kg)	0.5
23.	Insurance (%)	6
24.	Cost of medicines and vaccines	
	Weaner/fattener (Rs./month)	2
	Adults (Rs./month)	4
25.	No. of labourers required	1
26.	Labourers wages (Rs. per month)	900
27.	No. of piglets sold per sow per farrowing (2 months old)	4
28.	No. of fatteners sold per sow per farrowing (8 months old)	4
29.	Sale price of piglet	450
30.	Avg. wt. of fattener (kg.)	80
31.	Sale price of fattener	1360
32.	Income from manure	
	Weaner/fattener (Rs./month)	2
	Adults (Rs./month)	5
33.	No. of gunny bags per ton to feed	13.3
34.	Income from gunny bags (Rs./bag)	6
35.	Depreciation on sheds (%)	5
36.	Depreciation on equipments etc. (%)	10
37.	Margin money (%)	15
38.	Interest rate (%)	13.5
39.	Repayment period (years)	5
40.	Grace period (years)	1

Annexure – VId
ECONOMICS OF PIG FARMING - HERD PROJECTION CHART

Year	Month	Breeding Stock		No. of piglets born	No. of Lactating piglets $	No. of fatteners		Sale of	
		Ist Batch	IInd Batch			3-5 months $	6–8 months $	Piglets	Fatteners
I	1	G	-		-	-	-		
	2	G	-		-	-	-		
	3	G	-		-	-	-		
	4	P	G		-	-	-		
	5	P	G		-	-	-		
	6	P	G	55	-	-	-		
	7	P	P	55	-	-	-	20	
	8	L	P		45	-	-*	-	
	9	L	P	55	45	-	-	-	
II	10	P	P	55	-	20	-	20	
	11	P	L		45	20	-	-	
	12	P	L	55	45	20	-	-	
	13	P	P		-	20	20		
	14	L	P		45	20	20		

Annexure – VId (Contd.)

Year	Month	Breeding stock		No. of piglets born	No. of Lactating piglets $	No. of fatteners		Sale of	
		Ist batch	IInd batch			3-5 months $	6-8 months $	Piglets	Fatteners
	15	L	P	55	45	20	20	20	20
	16	P	P		-	20	20	-	
	17	P	L	55	45	20	20	-	
	18	P	L	55	45	20	20	20	20
	19	P	P		-	20	20	-	
	20	L	P	55	45	20	20	-	
	21	L	P	55	45	20	20	20	20
	22	P	P		-	20	20	-	
	23	P	L	55	45	20	20	-	
	24	P	L	55	45	20	20	20	20
III	So on..								

G–Growing period ; P–Pregnancy period ; L.–Lactating period.

$ No. of piglets and fatteners of different age groups for working out economics were taken after considering the mortality as it occurs usually at an early age in pigs.

Closing stock and it's value

1. Breeding stock (10+1) of original value.
2. One batch of 2 months old piglets: Sale value of piglets.
3. One batch of 5 months old : 60% of sale price of fattener.

Annexure – VIe

ECONOMICS OF PIG FARMING – CASHFLOW ANALYSIS

Sr.	Particulars	I	II–IV	V
I.	**Costs :**			
1.	**Capital Cost ***	94600	0	0
2.	**Recurring Costs :**			
	a) Breeder feed cost	18922	21499	21499
	b) Piglet feed cost	5400	10800	10800
	c) Fattener feed cost	4185	39060	39060
	d) Insurance cost	936	936	936
	e) Labour wages	10800	10800	10800
	f) Cost of medicines etc.			
	for breeder stock	468	528	528
	for weaner/fatteners	480	1680	1680
	g) Misc. expenses			
	for breeder stock	468	528	528
	for weaner/fatteners	480	1680	1680
	Total costs	136739	87511	87511
II	**Benefits :**			
1.	Sale of piglets	18000	36000	36000
2.	Sale of fatteners	0	108800	108800
3.	Sale of gunny bags	1276	3290	3290
4.	Sale of manure			
	From breeder stock	585	660	660
	From young stock	480	1680	1680
5.	Depriciated value of			
	a) Sheds	0	0	47625
	b) Equipment	0	0	7750

6.	Value of closing stock	0	0	48720	
	Total benefits	20341	150430	254525	
	Discount factor @15%	0.870	1.985	0.497	
	Discounted costs @ 15%	118903	173744	43508	336156
	Discounted benefits @ 15%	17688	298666	126544	442898
III	**NPW**	**Rs.**	**106742**		
IV	**BCR**		**1.32:1**		
V	**Net benefits :**	–116398	62920	167015	
	Discount factor @ 50%	0.667	0.938	0.132	
	Discounted net benefits @ 50%	–77599	59036	21994	3431
VI	**IRR**	**>**	**50%**		

* Excluded the capitalised amount on feed, insurance, labour wages, medicine cost and misc. expenses.

Annexure–VIf

ECONOMICS OF PIG FARMING – REPAYMENT SCHEDULE

Bank loan (Rs.)	116200
Interest (%)	13.5
Capital Recovery Factor	0.339693

Rupees

Year	*Income*	*Expenses*	*Gross Surplus*	*Equated Annual Instalment$*	*Net Surplus*
I	20341	o*	20341	11768	8573
II	150430	87511	62920	39471	23447
III	150430	87511	62920	39471	23447
IV	150430	87511	62920	39471	23447
V	150430	87511	62920	39471	23447

*Capitalised.

$During first year only interest will be recovered.

Note : Average loan period in first year is considered as 9 months for working out interest amount.

Annexure – VII

SPACE REQUIREMENT OF PIGS

Type of animals	*Floor space requirement (Sq.mt. per animal)*		*Maximum number of Animals per pen*
	Covered Area	*Open Paddock*	
Boar	6.0 – 7.0	8.8 – 12.0	Individual pens
Farrowing sow	7.0 – 9.0	8.8 – 12.0	Individual pens
Fattener (3 – 5 months old)	0.9 – 1.2	0.9 –1.2	30
Fattener (above five months)	0.3 – 1.8	1.3 – 1.8	30
Dry sow/gilt	1.8 – 2.7	1.4 – 1.8	3 –10

Annexure – VIIIa

NUTRIENT REQUIREMENT OF BREEDING STOCK

Type *Liveweight (kg.)*	*Breed Gilts* *110–250*	*Lactating gilts and sows* *140–250*	*Young boars & adult boars* *110–250*
Energy and protein			
DE (Mcal/kg)	3.3	3.3	3.3
ME (Mcal/kg)	3.17	3.17	3.17
Crude Protein (%)	14	15	14
Inorganic nutrients (%)			
Calcium	0.75	0.75	0.75
Phosphorus	0.5	0.5	0.5
Salt	0.5	0.5	0.5

Annexure – VIIIb

NUTRIENT REQUIREMENTS OF GROWING PIGS

Typeq Liveweight (kg) Daily gain (kg)	Weaning 5–12 0.3	Growing 12–50 0.5	Finishing 50–100 0.6
Energy and protein			
DE (Mcal/kg)	3.5	3.5	3.3
ME (Mcal/kg)	3.36	3.36	3.17
Crude protein (%)	22	18	14
Inorganic nutrients (%)			
Calcium	0.8	0.65	0.50
Phosphorus	0.6	0.50	0.40
Sodium	--	0.10	–
Chlorine	–	0.13	–

10

Plant Economics of Pig Farming

Rated Plant Capacity = 3000.00 NOS/Day

LAND & BUILDING

(Land Area 4000 Sq.mt.) = Rs. 49,96,000.00

PLANT & MACHINERY

1. Food Trags Bigger and Smaller Size
2. Bathing Tanks Bigger Size
3. Drinking Water Baths
4. Cold Air Blower & Air Cooler
5. Tube Well
6. Cleaning and Washing Equipment
7. Miscellaneous Equipments

FIXED CAPITAL

1.	Land and Building	Rs.	49,96,000.00
2.	Plant and Machinery	Rs.	9,35,000.00
3.	Other Fixed Assets	Rs.	6,25,000.00
	TOTAL	Rs.	65,56,000.00

RAW MATERIALS

1. Feeding
2. Cost of Feeding
3. Cost of Medicine

TOTAL WORKING CAPITAL/ANNUM

1.	Raw Material	Rs.	30,94,400.00
2.	Salary and Wages	Rs.	10,90,200.00
3.	Utilities and Overheads	Rs.	8,60,000.00
	TOTAL	Rs.	50,44,600.00

TOTAL CAPITAL INVESTMENT

Total Fixed Capital	Rs.	65,56,000.00
Total Working Capital for 1 Month	Rs.	4,20,383.00
TOTAL	Rs.	69,76,383.00

TURN OVER/ANNUM

1.	By Sale of 3000 Pig Weaner		
2.	Sale of Manure		
3.	Sale of Replace Stock 60 Pigs		
	TOTAL	Rs.	77,76,000.00

PROFIT SALES RATIO = Profit / Sales x 100 = 16%

RATE OF RETURN = Operating profit / T.C.I x 100 = 17%

BREAK EVEN POINT (B.E.P) = 66%

11

Plant Economics of Pig Farming

Rated Plant Capacity = 9920 Nos./Annum

LAND AND BUILDING

(Land 20 Acres) =	Rs.	9,75,000.00

PLANT & MACHINERY

1. Pots, buckets, Basinus, spades Daos etc.
2. Refrigerator 265 lits capacity with one stabliser
3. Miscellaneous items and other water and other water tubewells etc

FIXED CAPITAL

1.	Land and Building	Rs.	9,75,000.00
2.	Plant and Machinery	Rs.	30,000.00
3.	Other Fixed Assets	Rs.	4,73,000.00
	TOTAL	Rs.	14,78,000.00

RAW MATERIALS

1. Feed concentrate
2. Feed concentrate for piglets
3. Medicines and vaccines

TOTAL WORKING CAPITAL/ANNUM

1.	Raw Material	Rs.	29,56,800.00
2.	Salary and Wages	Rs.	3,25,000.00
3.	Utilities and Overheads	Rs.	30,500.00
	TOTAL	Rs.	33,12,300.00

TOTAL CAPITAL INVESTMENT

Total Fixed Capital	Rs.	14,78,000.00
Total Working Capital for 1 Month	Rs.	2,76,025.00
TOTAL	Rs.	17,54,025.00

TURN OVER/ANNUM

1. By sale of piglets of the second batch (at age of 2 months)(4960 Nos.)
2. By sale of piglets of the third batch (at the age of 2 months) (4960 Nos.)
3. By sale of undamaged gunny bags 7700 Nos.
4. By sale of manure about 125 Tonnes

TOTAL	Rs.	49,95,600.00

PROFIT SALES RATIO = Profit / Sales x 100 = 28%

RATE OF RETURN = Operating profit / T.C.I x 100 = 80%

BREAK EVEN POINT (B.E.P) = 23%

12

Plant Economics of Piggery Meat Processing

Rated Plant Capacity = 1.68 MT/Day

LAND & BUILDING

(Land 2 Acres) = Rs. 41,00,000.00

PLANT & MACHINERY

1. Bathing tank with water spraying system
2. High Voltage electrode
3. Hoisting Chain and clamp
4. Knives (double edged type) and others cutting instruments
5. Blood collection Tank (S.S)
6. Scalding Tank
7. Automatic dehairing machine
8. Furnace with flame producing burner
9. Washing Tank
10. Electric circular revolving Knife
11. Refrigeration Chamber with machine
12. Small automatic cutting knivesvertical,horizontal
13. Meat mincer automatic
14. Automatic meat Extruder
15. Mixer (S.S.) complete with all accessories
16. Cooker for piggery (S.S)
17. Automatic Can weighing,filling and double seaming M/C
18. Sterlisation autoclave
19. Deep Freezer
20. Semi Automatic Carton Packaging machinery,
21. Boiler

22. Conveyer belt,Bucket type elevator etc
23. Other miscellaneous equipments as pipes,fitting,Pumps,valves, automatic process controlequipments
24. Laboratory Testing equipments etc
25. Miscellaneous,tools jigs,fixtures
26. Effluent treatment units for liquid solid and gases effluent

FIXED CAPITAL

1.	Land and Building	Rs.	41,00,000.00
2.	Plant and Machinery	Rs.	43,90,000.00
3.	Other Fixed Assets	Rs.	29,00,000.00
	TOTAL	Rs.	1,13,90,000.00

RAW MATERIALS

1. Pigs
2. Tin containers
3. Cartons
4. Miscellaneous chemical salts,oils and other preservatives

TOTAL WORKING CAPITAL/MONTH

1.	Raw Material	Rs.	10,04,500.00
2.	Salary and Wages	Rs.	2,61,750.00
3.	Utilities and Overheads	Rs.	1,71,000.00
	TOTAL	Rs.	14,37,250.00

TOTAL CAPITAL INVESTMENTS

Total Capital Investment	Rs.	1,13,90,000.00
Total Working Capital for 1 Month	Rs.	14,37,250.00
TOTAL	Rs.	1,28,27,250.00

TURN OVER/ANNUM

By sale of Processed Piggery Meat 504 MT Rs. 3,52,80,000.00

PROFIT SALES RATIO = Profit / Sales x 100 = 43%

RATE OF RETURN = Operating profit / T.C.I x 100 = 117%

BREAK EVEN POINT (B.E.P) = 25%

13

Suppliers of Plant and Equipments

BOILERS

Fluidtech Boilers Pvt. Ltd.
2703, Phase-IV, GIDC
Vatva, Ahmedabad,
Gujarat-382445
TEL : +91 (0) 79 25840105
FAX : +91 (0) 79 25840041
E-mail: info@fluidltd.com

Shri Indtex Boiler Pvt. Ltd.
Plot No. 1601, GIDC Est.,
Phase-I, Naroda
Ahmedabad, Gujarat-382330
TEL : +91 (0) 79 22823057
FAX : +91 (0) 79 22820047
E-mail : shiyug@wilnetonline.net

MJ Patel (India) Ltd.
118, P D'Mello Rd.
Nr. Carnac Bridge
Mumbai, Maharashtra-400009
TEL : +91 (0) 22 23422114
FAX : +91 (0) 22 23427410
E-mail: mjpil@vsnl.net.in

Kemtech International Ltd.
A-125, Shivalik, Malviya Ngr.
P.O., New Delhi,
NCT of Delhi-110017
TEL : +91 (0) 11 26688343
FAX : +91 (0) 11 26680899
E-mail:
info@kemtechinternational.com

LG Thermoflo Systems Pvt. Ltd.
Ground Flr., Paigah Plaza
Basheer Bagh
Hyderabad,
Andhra Pradesh-500029
TEL : +91 (0) 40 23242468
E-mail:
rk_lgindia@yahoo.com

Crupp Metals
Kh. No. 56/1
Mundka, Rohtak Rd.
New Delhi,
NCT of Delhi-110041
TEL : +91 (0) 11 25183085
FAX : +91 (0) 11 25183085

Isotex Corp.
3rd Flr., Laxmi Chmbrs.
Navjivan Rd. Ahmedabad,
Gujarat-380014
TEL : +91 (0) 79 27544321
FAX : +91 (0) 79 27541224
E-mail: isotex@icenet.net

Ambica Engg. Works
16/A, 1st Main,
Modi Hospital Rd.,
W.O.C Road, 2nd Stage.
Bangalore,
Karnataka-560086
TEL : +91 (0) 80 23495370
FAX : +91 (0) 80 23496120
E-mail:
ambica_e_w@vsnl.com

Vijay Engineering
2nd Flr., Gulshan Bldg.
99, Bhandari St.
Nr. Bombay Andheri
Transport, Masjid Bunder
Mumbai-400003
TEL: +91 (0) 22 23451045
FAX: +91 (0) 22 23451045
E-mail:
vijayengineering@indiainfo.com

Enmas Andritz Pvt. Ltd.
4th Flr., Guna Bldg.
Annexe, 443, Anna Salai
Teynampet
Chennai-600018
TEL: +91 (0) 44 24338050
FAX: +91 (0) 44 24322412
E-mail: eal@vsnl.com

FEED AND HAY KNIVES & BLADES

Paras Engineering Co.
Plot No. C-20/3
MIDC Indl. Area
Taloja, Navi
Mumbai, 410208
TEL : +91 (0) 22 27411212
FAX : +91 (0) 22 27411211
E-mail:
paras@shahgroup.biz

Raghavendra Innovative Industries
3-55, Gopal Puram

Hanmakonda,
Andhra Pradesh-506015
TEL : +91 (0) 871 2565669

DB Engineering Pvt. Ltd.
A-119, Phase-II
Okhla Indl. Area
New Delhi-110020
TEL : +91 (0) 11 26385177
FAX : +91 (0) 11 26386453
E-mail: sales@skberi.com

Spa Worldwide
209, Kale Ram Chmbrs.
2, East Guru Angad Ngr.
New Delhi-110092
TEL : +91 (0) 11 22048381
FAX : +91 (0) 11 22048381
E-mail:
spa@mantraonline.com

Apex Knives Pvt. Ltd.
132/134, S V Patel Rd.
Dongri, Mumbai-400009
TEL : +91 (0) 22 23791409
FAX : +91 (0) 22 23737707
E-mail:
mohit@apexknives.com

Seba Engineers Pvt. Ltd.
D-26, Indl. Area
Bulandshahr Rd.
Ghaziabad,
Uttar Pradesh-201009
TEL : +91 (0) 120 24703703
FAX : +91 (0) 120 24703703

LABORATORY EQUIPMENT & SUPPLIERS

Sunjay Technologies Pvt. Ltd.
A-32, Shri Ram Indl. Est.,
13, G D Ambedkar Rd.
Wadala, Mumbai-400031
TEL : +91 (0) 22 24118800
FAX : +91 (0) 22 24111540
E-mail:
info@sunjaytechnologies.com

Technico Laboratory Products
21/34, Poomagal St.
Ekkattuthangal
Chennai-600097
TEL : +91 (0) 44 22344965
FAX : +91 (0) 44 22344965
E-mail:
technico@md3.vsnl.net.in

Thermolab Scientific Equipments Pvt. Ltd.
218, Sterling Chmbrs.,
1st Flr., 56, Mogra Village
Andheri (E)
Mumbai-400069
TEL : +91 (0) 22 28353017
FAX : +91 (0) 22 28394562
E-mail:
madhav@bom3.vsnl.net.in

Bellstone Hi-Tech International
3755, Chawri Bazar
Delhi-110006
TEL : +91 (0) 11 23939951
FAX : +91 (0) 11 23939953
E-mail:
sales@bellstonehite.com

Erection & Instrumentaion Engineers
1001/1002, Span Trade
Centre Opp. Kochrab
Ashram, Nr. Paldi Circle
Ahmedabad-380006
TEL : +91 (0) 79 26579074
FAX : +91 (0) 79 26584099
E-mail: vindish@icenet.net

Amarchand & Company Pvt. Ltd.
56, Indl. Est.
Ambala Cantt.,-133006
TEL : +91 (0) 171 2699499
FAX : +91 (0) 171 2699806
E-mail:
accolab@accolab.com

Abhay Instruments & Controls
1307, 11, Thmn. Rd.
Vijay Ngr., Bangalore,
Karnataka-560040
TEL : +91 (0) 80 23302184

Tempo Instruments & Equipments (I) Pvt. Ltd.
1, Lamington Chmbrs., 394
Dr. Bhadkamkar Marg
Mumbai-400004
TEL : +91 (0) 22 23855674
FAX : +91 (0) 22 23855708
E-mail: tempo@vsnl.com

Sharmasons Sakova Instruments Pvt. Ltd.
E-175, Kavi Ngr. Indl. Area
Ghaziabad-201002
TEL : +91 (0) 120 24700090
FAX : +91 (0) 120 24702041
E-mail:
sakovas@mantraonline.com

Indian Instruments Manufacturing Co.
208, Bipin Behari Ganguly
St. (Bowbazar St.)
Kolkata-700012
TEL : +91 (0) 33 2411848
FAX : +91 (0) 33 2415970
E-mail: iimc@cal.snl.net.in

MEAT PROCESSING MACHINERY

Superfreeze India Ltd.
WZ-92-A, Raja Garden
Ring Rd., Delhi-110015
TEL : +91 (0) 11 25122667
FAX : +91 (0) 11 25101061
E-mail:
corporate@superfreezeindia.com

Wintech Taparia Ltd.
3rd Flr., Shreesh Chmbrs.,
25/1, Y N Rd.
Indore-452003
TEL : +91 (0) 731 2433950

FAX : +91 (0) 731 2430527
E-mail:
sales@wintechtaparia.com

Chengalva Engineers Ltd.
405, Shashank Residency, St. No. 11, Tarnaka
Hyderabad-500017
TEL : +91 (0) 40 27017397
FAX : +91 (0) 40 27018517
E-mail:
chenga@hd1.vsnl.net.in

MIXER

NSW India Ltd.
Plot No. 43, Sect. 34
EHTP, Gurgaon-122001
TEL : +91 (0) 124 26371106
FAX : +91 (0) 124 26372859
E-mail:
corporate@nswindia.com

Chemac Equipments Pvt. Ltd.
M J D'Souza Compound
Safed Pool, Sakinaka
Mumbai-400072
TEL : +91 (0) 22 28510777
FAX : +91 (0) 22 28516986
E-mail:
chemac@bom4.vsnl.net.in

Machinkraft Equipment
337, 3rd Rd., Sindh Hsg. Soc.
Aundh, Pune-411007
TEL : +91 (0) 20 25887464
FAX : +91 (0) 20 25880353
E-mail:
info@machinkraftequipment.com

Continental Equipment India Pvt. Ltd.
B-66, Phase-I
Okhla Indl. Area
New Delhi-110020
TEL : +91 (0) 11 26816297
FAX : +91 (0) 11 26815297
E-mail: info@ceipl.com

Shri Momai Fabricators Engineers
C/57, Lakulesh-II
B/h Sardar Est.
Ajwa Rd., Vadodara,
Gujarat-390019
TEL : +91 (0) 265 2565566
FAX : +91 (0) 265 2569816
E-mail:
iyer@shreemomai.com

Peenya Alloys Pvt. Ltd.
No. 248, 3rd Cross, 8th Main, Phase-III
Peenya Indl. Area
Bangalore-560058
TEL : +91 (0) 80 28394259
FAX : +91 (0) 80 28396536
E-mail:
jominsol@satyamn.net.in

Febchem Engineering Works
Plot No. A/2, Sect. A, Indl. Area, Sanwer Rd.
Indore-452003
TEL : +91 (0) 731 2720467
FAX : +91 (0) 731 2723087
E-mail:
febchem@yahoo.com

Bright Pharma Machinery
37-A, New Empire Indl. Est.
Kondivita Rd., J B Ngr.
Andheri (W) Mumbai,
Maharashtra-400059
TEL : +91 (0) 22 28214341
FAX : +91 (0) 22 28210190
E-mail:
brightpharma@hotmail.com

Acufil Machines
S.F. 120/2, Kalpatty P.O.
Coimbatore,
Tamil Nadu-641035
TEL : +91 (0) 422 2866108
FAX : +91 (0) 422 2866255
E-mail:
acufil@md4.vsnl.net.in

Seba Engineers Pvt. Ltd.
D-26, Indl. Area
Bulandshahr Rd.
Ghaziabad-201009
TEL: +91 (0) 120 24703703
FAX: +91 (0) 120 24703703

REFRIGERATOR

G-Tek Corporation
3, Mahavir Indl. Est.
Opp. Karelibaug
Telephone Exch.
Bahucharaji Rd.
Vadodara-390018
TEL : +91 (0) 265 2461912
FAX : +91 (0) 265 2460127
E-mail: info@gtek-india.com

Grabner International
5/30, West Patel Ngr.
New Delhi-110008
TEL : +91 (0) 11 25887300
FAX : +91 (0) 11 25732664
E-mail:
grabnerinternational@rediffmail.com

Verma Frost
403-A, Phase-II
Indl. Area, Chandigarh,
Chandigarh (UT)-160002
TEL : +91 (0) 172 2652078
FAX : +91 (0) 172 2389955

Baweja Electricals & Refrigeration Works
B-3/22, Nr. Raja Garden
Bus Stop, Rajouri Garden
New Delhi-110027
TEL : +91 (0) 11 25431948
FAX : +91 (0) 11 25101158

AMI Cooling System
9,10, Shreenathji Est.
Panna Est. Rd., B/h BOC
Gas Company, Rakhial
Ahmedabad-380023
TEL : +91 (0) 79 2779914
FAX : +91 (0) 79 2778661

Rinac India Ltd.
No. 5, "Saraswathi Nivas"
Main Channel Rd.
Saraswathipuram
Ulsoor, Bangalore,
Karnataka-560008
TEL : +91 (0) 80 25542929
FAX : +91 (0) 80 25511750
E-mail: rinac@vsnl.com

Friz-Tech Pvt. Ltd.
213/208, Navyug Indl. Est.
T J Rd., Sewri
Mumbai-400015
TEL : +91 (0) 22 24139593
FAX : +91 (0) 22 24155177
E-mail: friztech@vsnl.net

Anand Refrigeration Centre
32, Swarnambiga Layout
Nr. Ganga Hospital
Ramnagar, Coimbatore,
Tamil Nadu-641009
TEL : +91 (0) 422 2231980

LG Electronics India Ltd.
Plot No. 51, Udyog Vihar
Phase-II, Surajpur Kasna
Rd., Greater Noida,
Uttar Pradesh-201306
TEL : +91 (0) 120 24560900
FAX : +91 (0) 120 24560921

Prince Refrigeration Company Pvt. Ltd.
82E, Dr. Sudhir Bose Rd.
Kolkata,
West Bengal-700023
TEL : +91 (0) 33 24494686
E-mail:
prince@cal2.vsnl.net.in

STORAGE TANKS

Mellcon Engineers Pvt. Ltd.
B-297, Phase-I
Okhla Indl. Area
New Delhi-110020
TEL : +91 (0) 11 26816530
FAX : +91 (0) 11 26816573
E-mail: mellcon@vsnl.net
www.mellcon.com

Devson Steels
C-15, Bulandshahr Rd.
Indl. Area, Ghaziabad,
Uttar Pradesh-201003
TEL : +91 (0) 120 24700804
FAX : +91 (0) 120 24700003
E-mail:
rakesh@devsonsteels.com

Steelite Engineering Ltd.
103/5, Blackie House
Opp. GPO, V T
Walchand Hirachand Marg
Mumbai-400001
TEL : +91 (0) 22 2324093
FAX : +91 (0) 22 2610803
E-mail:
samir@steelitegroup.com

Dipesh Engineering Works
3, Sheroo Villa, 87
J P Rd., Andheri (W)
Mumbai-400053
TEL : +91 (0) 22 26336729
FAX : +91 (0) 22 26315507
E-mail:
kjpatel_71@hotmail.com

Pragati Engineering Works
Plot No. 114, New Chikhloli
MIDC, Ambernath,
Maharashtra-421505
TEL : +91 (0) 251 2687407
FAX : +91 (0) 251 2685054
E-mail:
shantaflaker@vsnl.com

Richard Engineering (Bom.) Pvt. Ltd.
1st Flr., Brij Cmplx., K B Rd.
Railway Flyover Junction
Ambarnath (W)
Ambernath,
Maharashtra-421501
TEL : +91 (0) 251 2892421
FAX : +91 (0) 251 2892418
E-mail: richsnow@vsnl.com

Machinkraft Equipment
337, 3rd Rd., Sindh Hsg. Soc.
Aundh, Pune,
Maharashtra-411007
TEL : +91 (0) 20 25887464
FAX : +91 (0) 20 25880353
E-mail:
info@machinkraftequipment.com

Aero Engineers
Plot No. 3419, GIDC,
Phase-IV, Vatva
Ahmedabad-382445
TEL : +91 (0) 79 25840720
FAX : +91 (0) 79 25841041
E-mail: aero@aeroengg.com

Coppral Agencies
7-1-246/2, Balkampet
Hyderabad,
Andhra Pradesh-500016
TEL : +91 (0) 40 23730549

Megha Polycon Pvt. Ltd.
2nd Flr., SBBJ Bldg.
SRCB Rd., Fancy Bazar
Guwahati, Assam-781001
TEL : +91 (0) 361 2510047
E-mail:
meghapolytop@hotmail.com

117. MANUFACTURE OF TIN CONTAINERS
118. MANUFACTURE OF STORAGE TANKS, PRESSURE VESSELS HEAT EXCHANGERS
119. MARK II HAND PUMPS
120. MARUTI WORKSHOP CUM - SERVICE STATION
121. MECHANICAL JACKS
122. METAL CUTTING DIE DESIGN
123. METAL CUTTING OF AND GRINDING WHEELS (ABRASIVE CUTTING WHEELS)
124. METAL HOOKS & CLIPS
125. METAL SEPARATION (COPP-ER , TIN, LEAD) FROM SPENT WASH ACID
126. METALLIC RING JOINTS
127. METALLIC GASKET (SPIRAL WOUND)
128. MICROVEE & ABSOLUTE FILTER
129. MICROWAVE OVEN
130. MILD STEEL INGOTS
131. MINI STEEL PLANT
132. MODERN VEHICLE WORKSHOP
133. MOPED
134. M.S.HINGES
135. M.S.INGOT AND HR. STEEL STRUCTURALS
136. M.S.INGOT BY INDUCTION FURNACES
137. M.S.PIPES
138. M.S. WELDING ELECTRODE
139. MUFFLERS & SILENCERS FOR THREE WHEELERS
140. NAIL CUTTER WITH FILER & MANICURE
141. NICHROME WIRE
142. NICKEL LINED SCREENS
143. NON-FERROUS ALLOY ROLLING
144. NON-FERROUS FORGING
145. NON-FERROUS FOUNDRY
146. NON PRESSURE
147. INCANDESCENT LAMP
148. NUMBER COMBINATION LOCKS FOR LUGGAGES
149. NUTS & BOLTS
150. PAPER COATED ALUMINIUM AND COPPER WIRE
151. PETROMAX CONTAINER
152. PHOTO ETCHING OF STAINLESS STEEL PLATES
153. PIPE GALVANIZING PLANT
154. PISTON RING-AUTOMOBILE
155. PLANT PROTECTION EQUIPMENTS
156. PLATINUM LABORATORY APPARATUS
157. PRESSURE COOKER & ALUMINIUM UTENSILS
158. PRESSURE COOKER (ALUMINIUM)
159. PRESSURE DIE CASTING
160. PRINTED ALUMINIUM COLLAPSIBLE TUBES
161. PRINTED TIN CONTAINERS
162. PRINTING PRESS (CYLINDER MACHINE)
163. PUMPS FOR CHEMICAL INDUSTRY (SPECIAL)
164. RAILWAY SLEEPERS (M.S.)
165. RAZOR TWIN BLADE
166. R.C.C SPUN PIPES
167. RECONDITIONING OF M.S. DRUMS/BARRELS
168. RE-ROLLING COPPER AND BRASS SHEET AND RODS
169. RE-ROLLING MILLS
170. RESIN COATED SAND
171. RESIN CORED SOFT SOLDER WIRES
172. ROLLING MILL (BY INDUCTION FURNACE) & MANUFACTURE OF BARS, ANGLES, SQUARES, TUBES AND OTHERS
173. ROLLER BEARING & FORGING
174. ROLLING OF STAINLESS STEEL PATTA
175. RUBBER INSULATED PLIERS (HAND TOOLS)
176. RUBBING COMPOUND FOR AUTOMOBILES
177. SCIENTIFIC LABORATORY EQUIPMENTS
178. SCOOTER ASSEMBLING
179. SECONDARY LEAD EXTRAC -TION BY SCRAP BATTERY PLATES,PIPES & SHEET
180. SEAMLESS M.S. TUBES & PIPES
181. SELF TAPPING STEEL SCREW
182. SEWING NEEDLES
183. S.G. IRON & ALLOY STEEL
184. SHEET METAL PRODUCTS, (FERROUS/NON-FERROUS)
185. SHIP/MARINE CONTAINER
186. SHOCK ABSORBERS
187. SHOE EYELETS
188. SHOT AND GRITS BY AUTOMIZATION PROCESS
189. SHOVELS
190. SILICO MANGANESE ALLOYS
191. SILENCERS (MUFFLERS) EXHAUST & TAIL PIPE FOR ALL TYPES OF VEHICLES
192. SINTERED BEARING
193. SINTERED BUSHES
194. SINTERED METAL PRODUCTS
195. SOFT AND HARD FERRITES
196. SPANNERS
197. SPARK PLUGS
198. STEEL PLANT (ELECTRIC ARC FURNACE BASED- EAF)
199. SPHERIODAL GRAPHITE CAST IRON
200. SPRAY DRYER
201. STAINLESS STEEL HINGES
202. STAINLESS STEEL SHEET ROLLING TO PRODUCE STAINLESS STEEL UTENSILS
203. STAINLESS STEEL UTENSILS
204. STAPLE PINS,PAPER PIN , GEM CLIPS ETC.
205. STEEL CASTINGS
206. STEEL CHAIN
207. STEEL FOUNDRY
208. STEEL FURNITURES AND ELECTRICAL APPLIANCES
209. STEEL PLANT (MINI)
210. STEEL RODS AND COILS FROM SCRAPS
211. STEEL ROLLING MILL
212. STEEL STRIPS (COLD ROLLED) SILICON WITH GRAIN RIENTED FOR ELECTRIC USE
213. STEEL TRANSMISSION LINE TOWERS & ROLLING MILL TO PRODUCE STEEL SECTION
214. STEEL FURNITURES AND ELECTRICAL APPLIANCES
215. STEEL WIRE DRAWING AND GALVANIZING
216. STEEL WOOL
217. SUBMERGED ARC WELDED PIPES
218. SUBMERSIBLE PUMP MANUFACTURING
219. SUPER ENAMELLED COPPER WIRE
220. THREE WHEELERS
221. TIE-ROD ENDS
222. TIN CONTAINERS
223. TOOLROOM AND SHEET METAL PRODUCTS
224. TRACTOR TRAILERS
225. TRANSMISSION POWER FITTING
226. TUBULAR POLES
227. SUPER ENAMELLED COPPER WIRES
228. TUBULAR POLES FOR ELECTRICAL TRANSMISSION (BY FABRICATION PROCESS)
229. TUBULAR POLES OF M.S. & HIGH TENSILE STEEL
230. VACUUM CLEANERS
231. VACUUM FLASK (STAINLESS STEEL)
232. VALVES FOR REFRIGERA TION AND AIR-CONDITION
233. VEHICLE WELDING & PAINTING
234. VENETION BLIND
235. WASHING MACHINES (AUTOMATIC & COMPUTERISED)
236. WATER CONTROLLER (AUTOMATIC)
237. WATCH STRAPS/CHAINS/ BRACELETS/BELT BRASS & STAINLESS STEEL
238. WATER COOLERS
239. WELDED WIRE MESH
240. WELDING ELECTRODES
241. WICK STOVES
242. WIND MILL WINDOW FRAME (FERROUS & NON-FERROUS)
243. WRIST WATCH

359. DI ETHYL OXALATE
360. DI METHYL ORTHOPHTHALATE
361. DI-METHYL PHTHALATE
362. DICLOFENAC GEL
363. DIOCTYL PHTHALATE (DOP)
364. DI PHENYL GLYCERINE
365. DI PHENYL OXIDE
366. DINITRO-CHLORO BENZENE
367. DISTILLED WATER
368. DODECYL BENZENE SULPHONATE
369. DUSTLESS CHALK
370. EDTA & ITS SALTS
371. ELECTROLESS NICKEL PLATING ON PLASTICS
372. ENDOSULFAN
373. EPOXY RESIN BASED COMPOUND
374. ETHYL ACETATE
375. EHTYL ETHER
376. ETHYL ALCOHOL (POTABLE LIQUOR)
377. ETHYL HEXANOL
378. EXTRACTION OF ESSENTIAL OILS BY SUGAR CRITICAL FLUID (CARBON DIOXIDE) METHOD FROM FLOWERS, HERBAL & SPICES
379. FERRIC ALUM
380. FERRO CHROME LIGNO SULPHONATE
381. FERRO MANGANESE
382. FERRO SILICONE
383. FERRO VANADIUM FROM VANADIUM SLUDGE
384. FERROUS SILICATE
385. FERROUS SULPHATE
386. FERTILIZER FROM ANIMAL BLOOD & LEATHER WASTE
387. FLUORESCENT TUBE LIGHT POWDER
388. FOAMED PVC COMPOUNDING & ITS PRODUCTS
389. FORMALDEHYDE
390. FRACTIONAL DISTILLATION OF D.M.O (DEMENTHOLIZED OIL)
391. FRACTIONAL DISTILLATION OF ESSENTIAL OIL & MEDICINAL PLANT EXTRACT
392. FRICTION DUST (LIQUID & POWDER) FROM CNSL
393. FRUIT FLAVOURS
394. FURFURAL FROM RICE HUSK
395. GARLIC ACID
396. GASKET SHELLAC COMPOUND
397. GEAR OIL
398. GIBBERELLIC ACID
399. GLASS PUTTY
400. GLYCERINE
401. GOSSYPOL (POLY PHENOL) FROM COTTON SEED OIL
402. GUAR GUM POWDER
403. HAIR FIXER (HAIR GEL TYPE)
404. H - ACID
405. HENNA PASTE MAKING
406. HEPTAL DEHYDE
407. HIGH CARBON FERRO CHROME
408. HYDRATED CALCIUM SILICATE BRICK
409. HYDRATED LIME FROM SEA SHELL
410. HYDROCHLORIC ACID
411. HYDRO FLUORIC ACID
412. HYDROGEN PEROXIDE (BY AUTO-OXIDATION PROCESS)
413. ICE PACKS (SOLUTIONS TYPE, WHITE GEL TYPE, VIOLET SEMI SOLID POLYMER TYPE)
414. IMPROVING DROP POINT PARAFFIN WAX FROM 45-50°C TO 75-80°C
415. INDUSTRIAL ALCOHOL
416. INTEGERATED COMPLEX OF EASTER & ALLIED PRODUCTS (D.O.P, D.B.P, ETHYL ACETATE WIRE ENAMEL & CABLE JELLY)
417. IMFL (WHISKY) & COUNTRY LIQUOR
418. IRON OXIDE FOR MAKING FERITTE
419. IRON SULPHIDE
420. JUTE BATCHING OIL
421. KESH KALA TEL (HAIR DYE LOTION) (VASMOL 33, BLACK NITE TYPE)
422. L-LYSINE MONOHYDROCHLORIDE
423. LACTIC ACID FROM WHITE SUGAR BY FERMENTATION PROCESS
424. LDPE GRANULES FROM VIRGIN (LDPE RESIN)
425. LEAD EXTRACTION FROM BATTERY SCRAP
426. LEAD OXIDE (A) LEAD MONOXIDE (B) LEAD TETRA OXIDE (C) GREYLEAD OXIDE
427. LIQUID GLUCOSE & ITS BY PRODUCTS
428. LIQUID OXYGEN BOTTLING PLANT
429. LIQUID SHOE POLISH
430. LIQUID FLOOR POLISH
431. LUBE OIL VISCOSITY IMPROVED FOR P.P.G/P.E.G.
432. LUBRICANTS ASHLESS 100% COMBUSTION
433. MAGNESIUM CARBONATE AND MAGNESIUM BICARMONATE
434. MAGNESIUM HYDROXIDE POWDER
435. MAGNESIUM SILICATE
436. MAGNESIUM SULPHATE
437. MALACHITE GREEN
438. MALEIC ANHYDRIDE
439. MANGANESE SULPHATE
440. MANUFACTURING OF CARBON MONO-OXIDE WATER GAS
441. MENTHOL BOLD CRYSTALS FROM FLAKES
442. MENTHOL CRYSTAL & MENTHA OIL
443. MERCURIC OXIDE
444. METAL PRE-TREATMENT CHEMICALS
445. METHANE GAS BY SODIUM ACETATE & SODA LIME
446. METHYL ACETYL RICINOLATE
447. METOL
448. METOL FROM HYDROQUINONE & METHYLAMINE
449. MICANITE
450. MICRO NUTRIENT MIXTURE
451. MINERAL WATER
452. MINERAL WATER AND PET BOTTILNG PLANT
453. MINERAL WATER IN BOTTLES, GLASS AND POUCHES
454. MINI CEMENT PLANT (BY ROTARY KILN PROCESS)
455. MIXED FERTILIZER
456. MOSQUITO & FLIES REPELLENT AGARBATTI (INCENSE STICKS)
457. MONO CHLORO ACETIC ACID
458. MONOCROTOPHOS (TECHNICAL)
459. MOSQUITO COIL
460. MOSQUITO COIL & MAT
461. MOSQUITO MAT
462. MOTHER TINCTURE & BIO CHEMIC MEDICINES
463. NAPHTHALENE & PHENYL (INTEGRATED UNIT)
464. NATURAL MINERAL WATER BY REVERSE OSMOSIS PROCESS
465. NICKEL PLATING BRIGHTNER (PRIMARY OR CARRIER BRIGHTNER & SECONDARY BRIGHTNER)
466. NICKEL SULPHATE
467. NICOTINE FROM TOBACCO WASTE
468. NICOTINE SULPHATE FROM TOBACCO WASTE
469. NITRO BENZENE
470. NITRO CELLULOSE SANDING SEALER/LACQUER
471. NITRO MUSK
472. NITROGEN & OXYGEN GAS PLANT
473. NON-IONIC SURFACTANT (WETTING AGENT)
474. NO-CARB PASTE
475. OCTANOL
476. ORTHO NITRO PHENOL
477. OXALIC ACID FROM MOLASSES
478. OXALIC ACID FROMRICE HUSK
479. OXALIC ACID FROM TREE BARK
470. OXALIC ACID FROM WASTE VEGETABLES
471. OXYGEN GAS PLANT
472. OXYGEN GAS PLANT (AIR SEPERATION METHOD)
473. PARA-AMINO BENZOIC ACID
474. PARA-AMINO PHENOL
475. PARA TOLUENE SULPHONIC ACID
476. PECTIN FROM RAW PAPAYA
477. PERFUME (LEMON & OTHERS)

CRUDE NAPHTHA
570. TRIMETHYL AMMONIUM CHLORIDE
571. TRIPHENYL PHOSPHITE (T.P.P)
572. ULTRAMARINE BLUE (LIQUID)
573. UREA FORMALDEHYDE & MELAMINE FORMALDEHYDE POWDER
574. UNDECYLINIC ACID
575. VINYL ACETATE MONOMER
576. VITAMIN C
577. VITAMIN E
578. WASTE WATER TREATMENT PLANT FOR INDUSTRIAL SECTOR IN INDIA (ONLY MARKER SURVEY)
579. WIRE DRAWING LUBRICANT
580. WIRE ENAMEL
581. XANTHATES
582. YELLOW DEXTRIN
583. ZINC CHLORIDE
584. ZINC OXIDE
585. ZINC PHOSPHATING BY COLD PROCESS
586. ZINC SILICATE
587. ZINC STEARATE
588. ZINC SULPHATE
589. ZINC SULPHATE MONOHYDRATE
590. ACID BLACK
591. AURAMINE'O'
592. AZO DYES STUFF
593. DYE & DYE INTERMEDIATE
594. DYE INTERMEDIATES
595. MALACHITE GREEN
596. METHYLENE BLUE
597. PHTHALOCYANINE BLUE
598. PHTHALOCYANINE BLUE & GREEN
599. REACTIVE DYES
600. SULPHUR BLACK DYE
601. VAT DYES
602. WATER CHILLING PLANT
603. WADDING OIL (100%) FOR WADDING OF COTTON HOSIERY CLOTH IN THE DYEING PROCESS
604. WHITE OIL
605. WAX FLOOR POLISH

ELECTRICAL, ELECTONICS, COMPUTERS AND INFOTECH/IT PROJECTS

606. AIR CONDITIONING
607. ALUMINIUM ALLOY CONDUCTOR
608. ALUMINIUM CABLE
609. ALUMINIUM ELECTROYTIC CAPACITORS
610. AUDIO CASSETTE ASSEMBLING & RECORDING
611. AUDIO CASSETTES DUPLICATING RECORDING
612. AUDIO CASSETTES & AUDIO STUDIO
613. AUDIO CASSETTES PLANE & RECORDED
614. AUDIO MAGNETIC HEADS
615. AUDIO MAGNETIC TAPE
616. AUDIO/VIDEO CASSETTES
617. AUTO BULB/LAMPS
618. AUTOMATIC VOLTAGE STABILIZER
619. BATTERY PLATES
620. BACK OFFICE (Rs. 5000/-)
621 B/W TV & COMPUTER MONITOR PICTURE TUBE
622. BREAD BOARDS
623. CALL CENTRE
624. CAMERA (35 MM)
625. CAPACITORS
626. CARBON ELECTRODE USED FOR BATTERY CELL
627. CARBON POTETIOMETERS
628. CEILING FAN
629. CERAMIC INSULATOR
630. CHILDREN INFOTECH TRAINING INSTITUTE (Rs. 5000/-)
631. CHOKE AND PATTI
632. CHOKE AND STARTER
633. CHOKE USED FOR FLUORESCENT LAMPS
634. COLOUR TELEVISION
635. COLOUR AND BLACK & WHITE TELEVISION
636. COMPACT DISC
637. COMPACT DISC PLAYER (AUDIO/VIDEO)
638. COMPUTER EDUCATION INSTITUTE
639. COMPUTER ASSEMBLY
640. COMPUTER HARDWARE
641. COMPUTER KEYBOARD
642. COMPUTER PERIPHERALS
643. COMPUTER PRINTERS
644. COMPUTER RIBBON
645. COMPUTER RIBBON REINKING OR REFILLING
646. COMPUTER STATIONERY
647. COMPUTER STATIONERY & IMPORTED HARDWARE PARTS
648. COMPUTER TERMINALS
649. COMPUTERISED WASHING MACHINE (AUTOMATIC)
650. COMPUTER SOFTWARE
651. CONDENSER FOR MOTOR USING MPP FILM

652. CONTENT DEVELOPMENT CENTRE (EOU) (Rs. 5000/-)
653 CONTROL PANEL BOARD
654. COOLING COIL FOR AIR CONDITIONERS
655. COPPER STRIP COIL FROM SCRAP
656. CORDLESS TELEPHONES
657. CYBER CAFE
658. CYBERKIOSK (Rs. 5000/-)
659. DATA PROCESSING CENTRE (Rs. 5000/-)
660. D.C. MICRO MOTORS
661. D.G.SETS
662. DISH ANTENNA AND CABLE T.V. NETWORK EQUIPMENT
663. DISTRIBUTION TRANSFORMERS & REPAIRS
664. DOMESTIC ELECTRICAL APPLIANCES- ROOM COOLER, WASHING MACHINE, WATER HEATER, ELECTRIC ROOM HEATER
665. E-SCHOOL
666. E-COMMERCE/BUSINESS
667. ELECTRIC ENERGY METER
668. ELECTRIC FANS
669. ELECTRIC HORN FOR AUTOMOBILE
670. ELECTRIC LAMP/GLS (INCANDESCENT LAMP)
671. ELECTRIC MIXER
672. ELECTRIC MOTORS UP TO 10 HP. REWINDING OF ALL TYPES OF MOTORS WATER PUMPS
673. ELECTRIC MOTOR WINDING (FOR FAN, MIXIES, ETC.)
674. ELECTRIC STEAM IRON
675. ELECTRICAL APPLIANCES
676. ELECTRICAL FIXTURES
677. ELECTRICAL STAMPING
678. ELECTROLYTIC CAPACITORS
679. ELECTROMAGNETIC RELAY
680. ELECTRONIC BALAST/CHOK
681. ELECTRICAL CHOKE
682. ELECTRONIC DIGITAL WATCHES
683. ELECTRONIC DIGITAL WEIGHING MACHINE
684. ELECTRONIC FIRE ALARM
685. ELECTRONIC GAS STOVE LIGHTERS
686. ELECTRONIC PRESSURE INDICATORS, ELECTRICALS, ELECTRONIC LIQUID LEVEL INDICATORS, ELECTRONIC TEMPERATURE INDICATOR, DIGITAL TACHOMETER
687. ELECTRONIC TELEPHONE INSTRUMENTS
688. ELECTRONIC TOYS
689. ELECTRONIC WATCHES & CLOCKS
690. FAX MACHINES
691. F.H.P MOTORS
692. FLOPPY DISKETTES
693. FLUORESCENT LAMP STARTER
694. FLUORESCENT TUBULAR

LAMPS WITH INTRODUCTION TO MERCURY VAPOUR LAMP
695. FRANCHISEE COMPUTER EDUCATION CENTRE (Rs. 5000)
696. GAS DETECTOR (L.P.G)
697. GENERATOR SET & PUMP SETS
698. GENERATOR (BATTERY OPERATED)
699. GIS SERVICE CENTRE (GEOGRAPHICAL INFORMATION SYSTEMS)
700. HARDWARE FITTING FOR TRANSMISSION LINE OVERHEAD LINE MATERIAL
701. HEADERS FOR TRANSISTOR ICS SEMI CONDUCTOR
702. H.T & L.T INSULATORS
703. H.T & M.V INDUSTRIAL CUBICAL SWITCH BOARD
704. INFORMATION MOVING DISPLAY (L.E.D TYPE)
705. INSURANCE CLAIM PROCESSING CENTRE (EOU) (Rs. 5000/-)
706. INTEGRATED CIRCUITS
707. INTERNET BASED STOCK TRADING (Rs. 5000/-)
708. INTERNET SERVICE PROVIDER (I.S.P.)
709. JELLY FILLED TELEPHONE CABLES
710. LAP TOP COMPUTERS
711. LEAD ACID BATTERIES
712. LEAD ACID BATTERY PLATES AND ASSEMBLING OF BATTERY
713. LEGAL TRANSCRIPTION & SECRETARIAL SERVICES CENTER (EOU)(Rs. 5000/-)
714. LIGHT EMITTING DIODES (L.E.D)
715. LINEAR ICS TRAINER KIT
716. LOUD SPEAKER
717. L.T TRANSFORMER REPAIRING
718. MEDICAL TRANSCRIPTION CENTRE
719. METAL FILM RESISTORS
720. METALLISED POLYPROPYLENE, POLYESTER FILM CAPACITOR
721. MICA BASE ELECTRONIC COMPONENTS
722. MICA PAPER WASTE PAPER FROM MICA WASTE
723. MICRO PROCESSORS TRAINER KITS BASED ON MICRO PROCESSORS
724. MINI COMPUTER (PERSONAL COMPUTER)
725. MINIATURE CIRCUIT BREAKER (M.C.B)
726. MINIATURE WATCH BATTERIES (BUTTON CELL)
727. MIXER/GRINDER (MIXI)
728. MONO CHROME COMPUTER MONITOR
729. MOTOR START ELECTROLYTIC CAPACITOR
730. MULTI LAYER P.C.B
731. MULTIPLE RELAY FOR LOW VOLTAGE
732. NEON INDICATOR
733. NEON SIGN MANUFACTURE
734. OPTICAL FIBRE CABLES
735. ON LINE SHOPPING MALL (Rs. 5000/-)
736. OPTO MECHANICAL & ELECTRICAL EQUIPMENTS
737. PHOTO COLOUR LAB
738. PICTURE TUBE (B/W)
739. PLAIN PAPER COPIER
740. PLASTIC FILM CAPACITORS
741. POLYESTER CAPACITORS
742. PORTABLE GENERATOR SET
743. PORTAL
744. PORTABLE TELEVISION(TV)
745. POWER CAPACITORS
746. POWER INVERTERS
747. POWER PLANT
748. POWER TRANSFORMERS UP TO 600 KVA
749. PRINTED CIRCUIT BOARD
750. PVC WIRES & CABLES
751. RE-CONDITIONING OF PICTURE TUBE
752. RESIN CAST CT & PT (1 KV)
753. SEMI CONDUCTOR DEVICE
754. SEMI CONDUCTORS FOR TRANSISTORS & DIODES
755. SETTING UP OF A VIDEO STUDIO
756. SMOKE DETECTORS
757. SOLAR CELLS
758. SOLAR MODULES
759. SOLAR PHOTO VOLTAIC SYSTEM
760. SOLDER FLUXES
761. SOLAR WATER HEATING PANELS
762. STEEL FURNITURE & ELECTRICAL APPLIANCES
763. STEREO AMPLIFIERS
764. STEREO CASSETTE RECORDERS/PLAYERS
765. STREET LIGHT FITTINGS SURGE SUPPRESSOR
766. TANTALUM CAPACITORS
767. TEFLON COATED ELECTRIC CABLE
768. TELEMEDICINE (DISTANCE HEALTH CARE) (Rs. 5000/-)
769. TELEPHONE CORD/CABLE
770. TELEPHONE (PUSH BUTTON TYPE)
771. TELEPHONE (PUSH BUTTON & CORDLESS)
772. TELEVISION (CTV & B/W)
773. TELEVISION (3-D)
774. TELEVISION DEFLECTION COMPONENTS
775. TELEVISION SIGNAL BOOSTERS
776. TELEVISION TUNERS
777. TRACTION BATTERIES
778. TRAINING INSTITUTE OF MEDICAL TRANSCRIPTION
779. TRANSFORMER FOR B/W TV
780. TRANSFORMER FOR VOLTAGE STABILIZER & E.H.T
781. TRANSMISSION POWER LINE FITTING
782. TRANSMISSION TOWER FITTING
783. TV AUDIO EQUIPMENT CABINETS & THEIR ASSEMBLING UNIT
784. UN-INTERRUPTED POWER SUPPLY (U.P.S)
785. VARIABLE FREQUENCY
786. VARIABLE VOLTAGE A.C. DRIVE
787. VIDEO CASSETTES (COMPLETE MANUFACTURING & ASSEMBLING)
788. VIDEO CASSETTES RECORDER (V.C.R)
789. VIDEO FILM STUDIO
790. VOLTAGE REGULATOR FOR AUTOMOBILES
791. VOLTAGE STABILIZERS
792. VOLTAGE STABLIZER & T.V. GAIN BOOSTER
793. WAX AND CHEMICAL COATED, BRAIDED TINSEL WIRE
794. WEBSITE DESIGN & E-MAIL REGISTERING
795. WELDING ELECTRODES
796. WIND ENERGY POWER PROJECT (10 MW)
796A WIRES (METAL)

FOOD, AGRO FOOD, PROCESSED FOOD, AGRO PLANTATION, CULTIVATION, FARMING, DAIRY/MILK, TOBACCO/PAN MASALA, BREWERY & DISTILLERY, EDIBLE OILS, EOU FOOD PRODUCTS AND ALLIED PRODUCTS

797. ALCOHOLIC BEVERAGES & VENEGAR FROM COCONUT WATER
798. ALCOHOL DRINKS FROM ETHYL ALCOHOL BY MIXING OF VARIOUS FLAVOURS
799. ANTI SCALE COMPOUND FOR ADDING INTO SUGAR JUICE BOILING
800. APPLE JUICE CONCENTRATED & DEHYDRATED FRUIT & VEGETABLES
801. ARTIFICIAL FISH MEAL FOR POULTRY FEED
802. AUTOMATIC BISCUIT PLANT
803. BABY CEREAL FOOD AND MILK POWDER
804. BACTERIA FOR CANE JUICE
805. BAKER'S YEAST
806. BAKERY INDUSTRY
807. BANANA PUREE
808. BEER & WINE
809. BEER PLANT (BREWERY)
810. BIDI
811. BIDI & CIGARETTE
812. BISCUIT PLANT

813. BREAD (AUTOMATIC PLANT)
814. BLACK PEPPER (SPICES)
815. BREAD AND BISCUITS
816. BREEDING FARM
817. BROILER CHICKEN
819. BUTTON MUSHROOM CULTIVATION & PROCESSING
820. CANNING OF RASAGULLAS IN METAL CANS
821. CAFFEIN FROM TEA WASTE
822. CANNING OF FRUITS & VEGETABLES
823. CASEIN AND BY-PRODUCTS
824. CASHEW FRUIT JUICE FROM CASHEW FRUIT APPLE
825. CASHEWNUT KERNEL EXTRACTION FROM CASHEWNUT FRUIT
826. CASHEWNUT SHELL LIQUID & KERNEL PROCESSING
827. CASHEW FENI
828. CATECHEU (BY CHEMICAL PROCESS)
829. CATTLE FEED FROM TAPIOCA
830. CHEWING & BUBBLE GUM
831. CHICKEN/SHEEP MEAT PROCESSING
832. CHOCOLATE (MILK)
833. CIGARETTE AND BEEDIES
834. COCOA BUTTER FROM COCOA MASS
835. COCONUT PRODUCTS & BY PRODUCTS PROCESS COMPLEX
836. COLD/SOFT DRINKS
837. COLLECTION OF MILK AND MILK MAKING POWDER
838. COLLECTION OF MILK AND PACKING IN POLYTHENE POUCH (1/2 KG.1 KG. & 2 KG.)
839. CONDENSED MILK (SWEETNED)
840. CONFECTIONERY INDUSTRY (TOFFEE & CANDY, SEMI-AUTOMATIC PLANT)
841. COUNTRY LIQUOR FROM MOLASSES
842. CIGARETTE
843. DAIRY FARM AND DAIRY (MILK) PRODUCTS (PASTEURISED MILK, BUTTER, GHEE, PANEER)
844. DAIRY FARM TO PRODUCE MILK & PACKING IN POUCHES (50%) & CAN(50%)
845. DAIRY FOR MILK PROCESSING
846. DAIRY FARM (BUFALO)
847. DAL (PULSE) MILL UNIT
848. DEHYDRATION OF CARROT & GARLIC
849. DEHYDRATION & CANNING OF FRUITS & VEGETABLES
850. DEHYDRATION OF FIGS
851. DEHYDRATION OF FRUITS AND VEGETABLES
852. DEHYDRATED ONIONS AND ONION POWDER
853. DRY ICE BY BREAKING OF AIR
854. DRYING OF RED CHILLIES, HALDI, DHANIA, AND GREEN PEAS
855. EGG POWDER (40,000 EGGS PROCESSING PER DAY)
856. EXPORT OF PROCESSED FOODS AND MARINE PRODUCTS
857. FISH CANNING & POUCHING
859. FISH FARMING (PRAWN & OTHER MARINE PRODUCTS)
860. FISH MEAL
861. FLOUR MILL AND MUSTARD OIL
862. FOOD COLOUR & ORASTED GROUNDNUT GRAM PEAS, ETC IN POUCHES
863. FROG LEGS PROCESSING
864. FROZEN MEATS PROCESSING
865. FOOD FLAVOURS (WHISKY) VODKA, GRAPE, BUTTER SCOTCH)
866. FOOD PRODUCTS COMPLEX (DEHYDRATED ONIONS, GARLIC POWDER & FLAKES, CATTLE FEED, TOMATO POWDER, TOMATO PRODUCTS, CANNED FRUITS & VEGETABLES, TOMATO PURE, GROUN NUT OIL, REFIND OIL, DEHYDRATED GRAPES, BANANA POWDER & WAFFERS)
867. FRUIT JUICE, JAM, JELLIES & ALLIED PRODUCTS
868. FRUIT JUICE, PICKLES PROCESSING AND CANNING
869. FRUIT JUICE MAKING & PACKING IN PLASTIC CONTAINER/POUCHES
870. FRUIT PROCESSING (JAM & JELLIES)
871. FRUIT PULP & JUICE CONCENTRATES
872. FRUIT & VEGETABLE DRYING (FREEZE DRYING METHOD)
873. GOAT & SHEEP FARMING
874. GRAM DALL/PULSE MILL
875. GRAPE DEHYDRATION
876. HARD BOILED CANDY (TOFFEE & CANDY)
877. HERBAL CIGARETTES
878. HONEY PROCESSING & PACKAGING
879. ICE CREAM OF DIFFERENT FLAVOURS
880. ICE CREAM STABILIZER
881. ICE MAKING PLANT
882. INDIAN MADE FOREIGN LIQUOR (I.M.F.L)
883. INSTANT FOOD (IDLI MIX, DOSA MIX, GULAB JAMUN MIX)
884. INSTANT FOOD (INSTANT FOOD & FAST FOOD PARLOUR)
885. INSTANT NOODLES
886. INSTANT TEA
887. INSTANT TEA FROM BLACK TEA
888. INTEGERATED STARCH BAKING POWDER/YEAST INDUSTRY
889. IODIZED SALT
890. IODIZED SALT (ORDINARY MOISTURE-LESS/FREE FLOWING IN PLASTIC BAGS AND CONTAINERS)
891. JAM, JELLIES, FRUIT JUICE & ALLIED PRODUCTS
892. KATHA AND CUTCH
893. KHANDSARI SUGAR
894. KHANDSARI SUGAR & IMFL
895. LACTIC ACID FROM WHITE SUGAR BY FERMENTATION PROCESS
896. LACTOSE & BY PRODUCTS PROCESSING FROM MILK
897. LIQUID GLUCOSE AND ITS BYE PRODUCTS
898. MACRONI AND VERMICILLI
899. MALT EXTRACTION FROM BARLEY
900. MANGO PROCESSING (MANGO PULP, JUICE & SLICES)
901. MAYUR BRAND TYPE CHEWING TOBACCO
902. MINERAL WATER
903. MILK PRESERVATION & MARKETING TO WHOLE SELLERS (INPOUCH PACKING BY UHT TECHNOLOGY)
904. MILK PROCESSING AND PACKAGING OF MILK PRODUCTS
905. MILK PRODUCTS (CASEIN, LACTOSE, GHEE & WHEY POWDER)
906. MILK TOFFEE
907. MINI FLOUR MILL
908. MINI SUGAR PLANT
909. MISRI (PEARL SUGAR CANDIES)
910. MODERN RICE MILL
911. MURABBA
912. MUSTARD OIL EXTRACTION & REFINING
913. MUTTON TALLOW
914. NAMKEEN INDUSTRY (BHUJIA, CHANA CHUR ETC.)
915. NAMKEEN & SWEETS
916. NON-BASMATI RICE FROM PADDY
917. OLEORESIN, ESSENTIAL OIL, DYES & POWDER OF SPICES
918. OLEORESIN EXTRACTION FROM DIPTERO-CARPUT TURMINATUS AND PINUS KHASYANA
919. OLEORESIN EXTRACTION FROM CHILLI
920. PAN MASALA AND POUCH

1059. BUFFALO HORN TIP, HOOF
1060. BANQUET HALL
1061. BUTTON
1062. CALCINED PETROLEUM (C.P) COKE
1063. CANVAS SHOES
1064. CANVAS SHOES, JUNGLE BOOTS
1065 CARBON BRUSH, BRUSH HOLDER & SLIP RING
1066. CEMENT SHEETS WITH COIR FIBRE & OTHER SEGMENTS
1067. CERAMIC TILES (GLAZED) BY DOUBLE FIRING/HEATING
1068. CHILDREN RECREATION CENTRE
1069. COAL BRIQUETTES
1070. COLD STORAGE & ICE
1071. COLD STORAGE FOR FRUITS AND VEGETABLES
1072. COMPRESSOR (HERMETIC) FOR AIR CONDITIONER
1073. COSTUME JEWELLERY/ IMITATION JEWELLERY
1074. DECORATIVE LAMINATED SHEET (SUNMICA)
1075. DIAMOND CUTTING & EXPORTS
1076. DRINKING STRAW FROM PROPYLENE
1077. ENGINEERING COLLEGE
1078. ENTERTAINMENT CLUB
1079. FASHION TECHNOLOGY INSTITUTE
1080. FAST FOOD PARLOUR
1081. FAST FOOD (INSTANT FOOD & FAST FOOD PARLOUR)
1082. FIRE EXTINGUISHERS (SODA ACID TYPE)
1083. FISH NET
1084. FLORICULTURE (CUT FLOWER ROSE) WITH GREEN HOUSE
1085. FLOOR COVERING SHEET
1086. FILM STUDIO (VIDEO)
1087. FLUSH DOOR, CHIP BOARD, WOOD WOOL & OTHER INSULATING BOARDS
1088. GAS DETECTORS OF L.P.G
1089. GEMS MANUFACTURING
1090. GLASS BOTTLE BY SCRAP
1092. GOAT & SHEEP FARMING
1093. GOLD ELECTROPLATING
1094. GOLD JEWELLERY (E.O.U)
1095. GOLD PLATED SILVER JEWELLERY & CUTLERY
1096. GRANITE SLAB AND TILES
1097. GREEN HOUSE
1098. HARD CHROMIUM PLATING
1099. HEALTH CLUB AND BEAUTY PARLOUR CUM HAIR SALOON WITH SONA BATH
1100. HEALTH CLUB CUM BEAUTY PARLOUR TRAININGINSTITUTE
1101. HORN TIP,HOOVE BUTTON
1102. HOSPITAL
1103. HOSPITAL (100 BEDS)
1104. HOSPITAL (200 BEDS)
1105. ICE MAKING PLANT
1106. IMITATION AND COSTUME
1107. JEWELLERY (NECKLACE, EARRINGS EAR TOP ETC.)
1108. INITMATE SCENT CHEMICALS FROM ALL TYPES OF FLAVOUR
1109. INVESTMENT CASTING
1110. JUTE, COIR, GLASS ROPE/ SUTLI
1111. JUTE TWINE
1112. L.P.G BOTTLING PLANT
1113. L.P.G REGULATOR (DOMESTIC PURPOSE)
1114. LAMINATED PARTICLE BOARD & HARD BOARD
1115. LAMINATED SAFETY AND TOUGHENED GLASS
1116. LEASING & HIRE PURCHASE
1117. LICHEN (CHARILA)
1118. MANILA ROPES
1119. MEDICAL COLLEGE
1120. MATCH UNIT FROM WAXED PAPER
1121. MEDIUM DENSITY FIBRE BOARD
1122. MELAMINE CROCKERY
1123. MINERAL WATER
1124. MINI CEMENT PLANT
1125. MIRROR SILVER, GOLDEN, PINK, BLACK & SMOKES
1126. MODERN ADVERTISING AGENCY WITH DTP & FILM STUDIO
1127. MOTEL/SMALL HOTEL
1128. MULTI COLOUR PRINTING
1129. MULTI STOREY COMMERCIAL COMPLEX ALONGWITH RESIDENTIALAND DELUX FLATS FOR FOREIGN TOURISTS & REVOLVING RESTAURENT AT THE TOP
1130. MULTIPLE LAMINATION INDUSTRY
1131. MUSHROOM GROWING AND PROCESSING (BY DEEP FREEZING METHOD)
1132. NEWS PAPER PRINTING
1133. NICKEL LINED INDUSTRIAL SCREEN
1134. OFFSET COLOUR PRINTING PRESS (SIX COLOUR)
1135. OFFSET PRINTING PRESS
1136. P.C.C ELECTRIC POLES
1137. PHOTOGRAPHIC DEVELOPER & FIXER
1138. PVC EXTRUSION PROFILE (DOOR & WINDOWS)
1139. POP-CORN
1140. PHOTO ETCHING OF S.S.
1141. PRE STRESSED CEMENT CONCRETE PIPES
1142. PILFER PROOF CAPS
1143. PILFER PROOF CAPS AND CROWN CAP
1144. POUCH MAKING & GRAVURE PRINTING (ROTO PRINTING)
1145. POULTRY & FISH FARMING (INTEGERATED UNIT)
1146. PRINTING OF TIN SHEETS
1147. READYMADE GARMENTS
1148. RESTAURANT
1149. ROSE PLANTATION AND ROSE OIL EXTRACTION
1150. ROTOGRAVURE PRINTING
1151. SAFETY BELTS
1152. SAFETY MATCHES
1153. SHOE LACES
1154. SILK COCOON CULTIVATION (GROWING OF SILK COCOON WARM)
1155. SILK SCREEN PRINTING FORMULATIONS FOR PLASTIC, PAPER & CLOTH
1156. SPIKENARD (JATAMANSI)
1157. STENCIL COATING SOLUTION
1158. SUGAR CANDY (MISRI)
1159. SUGAR CANE PLANTATION
1160. SUGAR CANE JUICE PRESERVATION
1161. SURAT ZARI
1162. SYNTHETIC SHOES & SOLES
1163. TEA PACKAGING INDUSTRY
1164. TENNIS BALLS
1165. THREE STAR HOTEL
1166. TRAVELLING AGENCY
1167. TRAYS, TROLLEYS FOR HOSPITAL WITH SCRATCHLESS COATING
1168. WAX EXTRACTS (TANNING POWDER)
1169. WIRE ROPE SLINGS
1170. WOOD WOOL SLAB
1171. WOODEN CANE FURNITURE WITH EXPORT POTENTIAL
1172. WOODEN FURNITURE
1173. WOODEN PANEL INCLUDING KILN SEASONING
1174. WOODEN FURNITURE
1175. WATCH CASE BUFFING
1176. WATCH DIAL
1177. WRIST WATCHES
1178. WATCHES (ELECTRONIC)
1179. ZEDOARY (KACHUR)

EDIBLE OILS, ESSENTIAL OILS LURICATIING OILS, GREASES, VEGETABLE OILS, WAXES, CAMPHOR, PERFUMES & PERFUMERY COMPOUNDS AND REFINED OILS ETC.

1180. AROMATIC PERFUMERY COMPOUND
1181. AEROSOL
1182. AGARBATTI PERFUMERY COMPOUND
1183. BEESWAX MANUFACTURE
1184. CAMPHOR
1185. CARDANOL FROM CASHEW NUT SHELL LIQUID
1186. CASTOR OIL
1187. CASTOR OIL DERIVATIVE OLEORESINS
1188. CHILLI OIL
1189. CITRONELLA OILS
1190. CLOVE OIL
1191. CONCENTRATE OF ROSE JASMINE & LILY ETC.
1192. COLOURED FLAME CANDLE

1193. CANDLES (SEMI-AUTOMATIC)
1194. CORN OIL (MAIZE OIL)
1195. DEHYDRATED CASTOR OIL
1196. EUGENOL FROM CINNAMON LEAF OIL
1197. EXTRACTION OF ESSENTIAL OIL (CARDAMON, JEERA, AJOWAN, GINGER OILS ETC. & PACKAGING OF GROUND SPICES)
1198. EXTRACTION OF JASMINE ESSENCE
1199. EXTRACTION OF ESSENTIAL OILS BY SUPER CRITICAL FLUID METHOD FROM FLOWERS, HERBS & SPICES
1200. EUCALYPTUS OIL
1201. EXTRACTION OF OIL FROM OIL SEED EXPANDEREXTRUSION TECHNOLOGY)
1202. FAT LIQUOR SULPHATED OIL
1203. FLAVOURS FOR FOOD
1204. GARLIC OIL & POWDER
1205. GINGER OIL, SANDALWOOD OIL AND NAGARMOTHA OIL
1206. GINGER OIL
1207. GINGER OIL & GINGER DUST
1208. INTEGERATED WAX COMPLEX
1209. IONONE FROM LEMON GRASS OIL
1210. JASMINE & LILLY FLOWER OIL
1211. LEMON GRASS OIL
1212. LIQUID PARAFFIN
1213. LUBE OIL & GREASE
1214. LUBRICATING OIL
1215. MENTHOL CRYSTALS
1216. MENTHOL OIL & CRYSTAL
1217. MICRO CRYSTALLINE WAX
1218. MUSTARD OIL (EDIBLE OIL)
1219. OIL FROM ARTEMISIA HERBS
1220. PALM OIL CRUSHING UNIT
1221. PAN MASALA
1222. PARAFFIN WAX
1223. PARAFFIN WAX FROM SLACK WAX
1224. REFINED OIL-SUN FLOWER OIL, GROUNDNUT OIL, STAFF FLOWER OIL & COTTON SEED OIL
1225. REFINED VEGETABLE OIL
1226. RECLAMATION OF USED ENGINE OIL
1227. RICE BRAN OIL (R.B.O)
1228. ROSE OIL EXTRACTION
1229. SMOKE LESS CANDLE
1230. SPICE OIL & OLEORESINS
1231. SOLVENT EXTRACTION PLANT (OIL CAKE BASED)
1232. SYNTHETIC ALMOND OIL
1233. SYNTHETIC TALLOW
1234. SYNTHETIC GHEE
1235. VEGETABLE OIL EXTRACTION & REFINING
1236. WADDING OIL (100%) FOR WADDING OF COTTON HOSIERY CLOTH IN DYEING PROCESS
1237. WAX EMULSION FOR TEXTILE
1238. WAX FLOOR POLISH

PAINT, ENAMEL, SOLVENTS, THINNERS, INKS & VARNISH

1239. ACRYLIC EMULSION PAINTS
1240. ALUMINIUM PAINT
1241. AUTOMOBILE PAINTS
1242. BALL POINT PEN REFILL INK
1243. BITUMINIOUS ROAD EMULSION
1244. BITUMEN
1245. BITUMINIOUS FELTS FOR WATER & DAMP PROOFING
1246. BITUMINIOUS ROAD EMULSION RAPID MEDIUM & SLOW SETTING
1247. BUFFING & POLISHING
1248. CEMENT PAINT FOR WHITE & GREY CEMENT

1249. DISPERSANT
1250. DRY DISTEMPER
1251. DRY DISTEMPER & CEMENT PAINT
1252. DUPLICATING INK BLACK FOR GESTNER DUPLICATOR
1255. EMULSION PAINTS (WATER BASED)
1256. HAMMERTONE PAINT
1257. INSULATING VARNISH & WIRE ENAMEL
1258. IRON OXIDE PIGMENTS
1259. INSULATING VARNISH (POLY VINYL BUTYRAL BASED, FFC GRADE)
1260. LIME COLOUR (CEMENT COLOUR)
1261. SYNTHETIC RED OXIDE FOR FLOORINGS
1262. SOLVENT & THINNERS
1263. MARKING INKS (WATER PROOF)
1264. METAL NAPHTHANATE (AS DRIER FOR PAINTS)
1265. N.C.PUTTY
1266. OFFSET PRINTING INK
1267. OIL BOUND DISTEMPER PAINT
1268. PAINT BRUSHES
1269. PAINT REMOVERS
1270. PAINT INDUSTRY
1271. PAINT & VARNISH
1272. PICTURE VARNISH
1273. POWDER COATING PAINT
1274. PRINTING INKS (OFFSET, FLEXO & ROTO GRAVURE)
1275. PRIMER PAINT & ENAMEL PAINT
1276. PUTTY & WATER PROOFING PAINT
1277. PRINTING INKS (FLEXO-GRAPHIC INK)
1278. PUTTY (METAL CASEMENT)
1279. RED OXIDE PRIMER (ANTI CORROSIVE)
1280. REFRECTORY PAINT (GRAPHITE BASED)
1281. SCREEN PRINTING INKS
1282. SILK SCREEN PRINTING INK FORMULATION FOR PLASTIC, PAPER, CLOTH
1283. STAMP & PAD INK
1284. STOVING PAINT
1285. SPIRIT SOLUBLE MALEIC RESIN
1286. TEXTURE PAINT
1287. THINNERS AND ITS ALLIED PRODUCTS
1288. TONER INK
1289. VARNISH THINNER (SOLVENT)
1290. VACUUM METALLIZING LACQUERS
1291. WOOD PRIMER

PHARMACEUTICAL, DRUGS, AYURVEDIC/HERBAL COSMETICS & MEDICINES, HOMOEOPATHIC MEDICINES, DISPOSABLE SYRINGE, DENTAL COLLEGE & FINE CHEMICALS ETC.

1292. AROMATIC PILLS
1293. ASPIRIN
1294. AYURVEDIC PAIN BALM OINTMENTS
1295. AYURVEDIC CHURAN AND TABLETS
1296. AYURVEDIC TABLETS (HAJMOLA TYPE)
1297. AYURVEDIC/HERBAL PHARMACY
1298. AYURVEDIC SHERBATS
1299. AYURVEDIC PRODUCTS
1300. BULK DRUGS (E.O.U)
1301. BLOOD BAGS
1302. CALCIUM GLUCONATE
1303. CAPSULE, TABLET & INJECTION WITH MODERN INSTRUMENTS
1304. CASHEW FENI
1305. CHLOROQUINONE PHOSPHATE (BULK DRUGS)
1306. CLINICAL THERMOMETER
1307. DENTAL CLINIC

1308. DENTAL COLLEGE
1309. DENTAL GRADE EUGENOL
1310. DEXTROSE MONOHYDRATE, LIQUID GLUCOSE
1311. DEHYDRATED ONION & ONION POWDER
1312. DEXTROSE POWDER (ANHYDROUS FROM STARCH)
1314. DEXTROSE SALINE SOLUTION
1315. DISTILLERY (I.M.F.L)
1316. DISPOSABLE NEEDLES FOR SYRINGES
1317. DISPRIN
1318. DISTILLED WATER
1319. EMPTY HARD GELATINE CAPSULES
1320. FILLING & PACKING OF CAPSULES
1321. GLUCOSE - D - POWDER
1322. GLYCERINE
1323. HERBAL COSMETICS
1324. HERBAL SHAMPOO
1325. HERBAL HAIR OILS (AYURVEDIC)
1326. HOMOE & BIO-MEDICINES WITH MOTHER TINCTURE
1327. HOMOEPATHIC MEDICINES
1328. HYPODERMIC NEEDLES
1329. I.V. FLUIDS
1330. IBUPROFEN
1331. INFUSION & TRANSFUSION SETS (I.V. SET)
1332. INJECTABLE FOR PHARMACEUTICALS
1333. INJECTION AMPOULES PACKAGING BOX
1334. INTEGERATED SURGICAL COTTON
1335. ICE MAKING PLANT
1336. ICE & COLD STORAGE
1337. LACTOSE & BY PRODUCTS PROCESSING FROM MILK
1338. LICHEN (JATAMANSI CHARILA)
1339. LIQUID GLUCOSE AND ITS BY-PRODUCTS
1340. L-LYSINE MONOHYDRO CHLORIDE
1341. LIQUID GLUCOSE FROM POTATOES
1342. MEDICAL COLLEGE
1343. NICOTINE FROM TOBACCO WASTE
1344. OINTMENT-AYURVEDIC (YELLOW & WHITE)
1345. PHARMACEUTICAL AND FOOD GRADE GELATINE
1346. PHARMACEUTICAL INDUSTRY (TABLETS, CAPSULES, LIQUID, GEL, OINTMENT POWDER INJECTABLE)
1347. PHARMACEUTICAL UNIT (EOU) WITH FORMULA TIONS, INJECTABLES, ETC. PYRIDINE & DERIVATIVES
1348. SALT (IODIZED SALT)
1349. SALINE & INJECTION WATER
1350. SALINE WATER & DEXTROSE SOLUTION (I.V.FLUID IN PLASTIC BOTTLES)
1351. SORBITOL
1352. STARCH (MAIZE)
1353. SURGICAL ADHESIVE PLASTER
1354. SURGICAL BANDAGES
1355. SURGICAL GLOVES
1357. SURGICAL COTTON
1358. SURGICAL COTTON AND BANDAGE
1359. SYNTHETIC CAMPHOR POWDER
1360. TABLETS & CAPSULES
1361. TISSUE CULTURE
1362. TINCTURE FROM RECTIFIED SPIRIT
1363. TRIMETHOPIME
1364. VETERENARY MEDICINES (ONLY FORMULATIONS)
1365. VITAMIN 'C', SORBITOL' ANHYDROUS DEXTROSE, STARCH

PULP, PAPER, STRAW/GREY BOARD, KRAFT PAPER, PACKAGING, PAPER CARRY BAGE, PAPER FROM AGRO WASTE & STATIONERY ETC.

1366. AMMONIA PAPER
1367. AUTOMATIC BOOK BINDING
1368. ADHESIVE (FEVICOL TYPE)
1369. ALL PIN & GEM CLIPS
1370. CARBON PAPER
1371. CARD BOARD
1372. CARBON LESS PAPER
1373. CELLPHANE PAPER
1374. COMPUTER FORMS & SECURITY PRINTING PRESS
1375. CORRUGATED BOARD & BOX (PRINTED & LAMINATED)
1376. CORRUGATED BOARD AND BOXES FROM FROM CARD BOARDS
1377. CORRUGATED CARTONS FROM PLAIN PAPERS
1378. CORRUGATRED PACKING & MATERIAL (BULB & TUBES PACKING)
1379. DESK TOP PUBLISHING
1380. DEFOAMING AGENT FOR PAPER PLANT
1381. DRINKING STRAW PAPER
1382. D.T.P CUM OFFSET PRESS
1383. EGG TRAYS
1384. EXERCISE NOTE BOOK, REGISTER AND PAD
1385. FLEXOGRAPHIC INK
1386. GREETING CARD BY OFFSET PRESS
1387. HAND MADE PAPER FILTER PAPER
1388. HARD BOARD
1389. INJECTION AMPOULES PACKAGING BOXES
1390. INSULATIING PAPER
1391. KRAFT PAPER
1392. KRAFT PAPER FROM BAGASSE
1393. LAMINATED PACKAGING PAPER WITH PRINTING
1394. LAMINATION & COATING ON PAPER
1395. M.G. PAPER FROM WASTE PAPER
1396. MILL PAPER FROM WASTE PAPER
1397. MILL BOARD FROM RICE & WHEAT STRAWS
1398. MILL BOARD FROM WASTE PAPER
1399. MINI PAPER PLANT FROM SISAL
1400. MINI PAPER PLANT
1401. MULTI WALL PAPER SACKS
1402. NEWS PAPER FOR CHILDREN
1403. NEWS PRINT PAPER FROM RICE STRAW BAGASSE
1404. NEWS PAPER CHILDRENS
1405. NEWS PRINT PAPER
1406. OFFSET AND TREADLE TYPE PRINTING PRESS
1407. MILL BOARD BASED ON RICE STRAW
1408. PAPER AND BOARD FROM STRAW
1409. PAPER & PAPER PRODUCTS
1410. PAPER ENVELOPES
1411. PAPER CARRY BAGS
1412. PAPER BOARD CARTON
1413. PAPER CONES FOR LOUD SPEAKERS
1414. PAPER FROM AKRA
1415. PAPER FROM BAGASSE WITH CORRUGATED BOARD & BOXES
1416. PAPER FROM BAMBOO
1417. PAPER GLASSES FOR BEVERAGES
1418. PAPER FROM RICE HUSK & WHEAT HUSK
1419. PAPER FROM TREE BARK, EUCALYPRUS WOOD
1420. PAPER PRODUCTS
1421. PAPER PLANT (WHITE WRITING & NEWS PAPER FOR PULP & WASTE PAPER)
1422. PAPER PLATES, PAPER GLASS
1423. PAPER CONES & TUBES
1425. PAPER PLANT WITH DTP & PRINTING & PUBLISHING UNIT
1426. PAPER BASED PHENOLIC SHEET
1427. PAPER CUP FOR ICE CREAM
1428. PAPER TUBES SPIRAL WINDING COMPOSIT CONTAINER
1429. PAPER LABLE FOR BEER BOTTLES
1430. PARTICLE BOARD & BLACK BOARD WITH SANDING & LAMINATION OF PARTICLE BOARD
1431. PARTICLE BOARD FROM RICE HUSK
1432. POUCH FILLING & MAKING FOR TOMATO SAUCE
1433. POUCH MAKING &

GRAVURE PRINTING
1434. PROCESSING OF PAPER FOR FEEDING IN COMPUTER
1435. PULP FROM BAMBOO & WOOD
1436. PAPER FILES
1437. PLAYING CARDS
1438. ROSIN SIZING AGENT (FOR PAPER PLANT)
1439. SAND PAPER
1440. SANITRY NAPKINS
1441. SILICON COATED PAPER
1442. STENCIL PAPER
1443. STRAW BOARD AND GREY BOARD
1444. STRAW BOARD & MILL BOARD
1445. STRAW BOARD AND PAPER BOARD
1446. TISSUE PAPER FACIAL
1447. TISSUE MOIST TOILETERY CLEANSING TISSUE AND RELATED PRODUCTS
1448. TISSUE PAPER ROLLS
1449. TETRA PACK FOR MILK PACKAGING, GHEE & OTHER LIQUIDS
1450. TOILET PAPER ROLLS
1451. TOILET PAPER & NAPKINS
1452. WAX COATED PRINTED PAPER
1453. WET FACE FRESHNER TISSUE
1454. WHITE WRITING & PRINTING PAPER
1455. WRITING & PRINTING PAPER (PAPER MILLS)

PLASTIC, B.O.P.P, ACRYLIC, DISPOSABLE PLASTIC PRODUCTS, PET PRODUCTS, P.V.C, H.D.P.E, P.P, L.D.P.E., P.U, A.B.S, THERMOFOARMING, MASTER BATCHES, & POLYMER AND RUBBER PRODUCTS, TYRE, TUBE, ADHESIVE, SHEET, COIR & MANY OTHERS

1456. ABS GRANULES FROM ABS SCRAPS FROM OLD T.V. CABINETS, WHITE GOODS ETC.
1457. ACRYLIC BEADS
1458. ACRYLIC COPOLYMER EMULSION
1459 ACRYLIC LATEX
1460. ACRYLIC SHEET
1461. ACRYLIC SHEET & MOULDED PRODUCTS
1462. ACRYLIC TEETH
1463. AUTO TUBES
1464. AUTO FLAPS FOR TRUCKS & BUSES
1465. AUTO TYRES & TUBES
1466. BABY BOTTLES (PLASTIC) WITH WHITESILICON RUBBER NIPPLES
1467. B.O.P.P. FILM
1468. B.O.P.P. PRESSURE SENSITIVE SELF ADHESIVE TAPE
1469. BABY NIPPLE (SILICON)
1470. BABY NIPPLE (BIG SIZE)
1471. BALLOON PLASTIC ADVERTISING
1472. BLISTER FILM PVC
1473. BLISTER PACKAGING & POUCH PAKAGING
1474. BLOW MOULDING PLASTIC CONTAINER
1475. COATING ON METALIZED POLYESTER FILM/METALISED PAPER/ ALIMINIUM FOIL
1476. COATING ON PLASTIC (ELECTROLYSIS) & GLASS
1477. COIR FOAM (RUBBERISED)
1478. COLOUR COATING ON PLASTICS
1479. COLOUR MASTER BATCHES FOR VARIOUS PLASTICS
1480. CYCLE TYRES AND TUBES
1481. DISPOSABLE PLASTIC CUPS GLASSES, ETC.
1482. DISPOSABLE PLASTIC SYRINGES & NEEDLES
1483. DISPOSABLE PLASTIC SYRINGES
1484. DOUGH MOULDING COMPOUND (D.M.C)
1485. EPOXY RESIN
1486. EXPANDED POLYESTRENE MOULDING (THERMOCOLE)
1487. F.R.P AUTO,SCOOTER ROOFS & CEILINGS
1488. F.R.P PRODUCTS (HELMET, WASHBASIN SHEETS, ROOFING SHEETS)
1489. FIBRE REINFORCED PLASTIC (HIGH PRESSURE MOULDING WITH SMC BMC AND DMC)
1490. FIELD RUBBER CONVERTED TO THE 60% LATEX RUBBER
1491. FLEXIBLE P.U. FOAM
1492. FORMALDEHYDE CROCKERY & OTHER ITEMS
1493. FORMALDEHYDE RESIN (UREA, PHENOL,MELEMINE)
1494. GASKET SHEET
1495. GUM BOTTLE (PVC)
1496. GLASS BEADS
1497. HDPE COATED PAPER SACK
1498. H.D.P.E. BAGS
1499. H.D.P.E. CONTAINERS (BLOW MOULDING)
1500. H.D.P.E. CONTAINER, POLY JARS BY INJECTION MOULDING (FOOD GRADE)
1501. H.D.P.E. JERRY CANS
1502. HDPE MANUFACTURING FROM ETHYLALCOHOL
1503. H.D.P.E PIPE AND FITTINGS
1504. H.D.P.E. PIPES
1505. H.D.P.E PRINTED BAGS
1506. H.D.P.E TWINES AND ROPES
1507. H.D.P.E. & L.D.P.E PIPES AND FITTINGS
1508. H.D.P.E/PP BOX STRAPPING
1509. H.D.P.E/PP WOVEN SACKS USING PLAIN LOOMS
1510. HOLOGRAM STICKERS-3D
1511. H.M BAG PLANT WITH PRINTING UNIT
1512. HAWAI CHAPPALS (RUBBER)
1513. I.V PLASTIC BOTTLE
1514. ICE CREAM CUP (PLASTIC)
1515. INJECTION MOULDED AUTO COMPONENTS
1516. INJECTION & BLOW MOULDED PLASTIC PRODUCTS
1517. INJECTION MOULDED PLASTIC PRODUCTS
1518. INTEGERATED COMPLEX ESTER'S & ALLIED PRODUCTS (D.O.P,D.B.P, & ETHYL ACETATE, BUYTL ACETATE, WIRE ENAMELS, JELLY CABLE COMPOUND)
1519. L.D.P.E. (LOW DENSITY POLY ETHYLENE) GRANULES FROM VIRGIN (L.D.P.E RESIN)
1520. INTEGERATED SURGICAL RUBBER GOODS INDUSTRY
1521. INJECTION MOULDED PLASTIC BALLS
1522. L.D.P.E FROM ETHYL ALCOHOL
1523. LDPE MOULDED PRODUCTS
1524. LAMINATION OF COEXTRUSION MULTILAYER FILM IN ROLL FORM
1525. LATEX RUBBER
1526. LATEX RUBBER CONDOM
1527. MASTER BATCHES (COLOURED, P.V.C, L.D.P.E, H.D.P.E, ETC.)
1528. MELAMINE FORMALHYDE RESIN
1529. MOULDED LUGGAGE
1530. MULTI-LAYER (3 LAYER BAGS)
1531. MULTI-LAYER (3 LAYER) FILM WITH LAMINATION & PRINTING
1532. OIL SEAL
1533. PET BOTTLE/CONTAINERS
1534. PET BOTTLES FROM PRE FORM
1535. PET BOTTLE & MINERAL WATER
1536. PET PRE-FORM FROM PET RESIN
1537. PET PRE-FORM CUM PET BOTTLES
1538. PET PRE-FORM PET BOTTLES CUM MINERAL WATERS
1539. POLY CARBONATE RESIN
1540. PTFE COMPONENTS
1541. PU/PVC SOLE FOR SPORT SHOE BY IMPORTED M/C.
1542. PVC BATTERY SEPARATOR
1543. P.V.C COMPOUNDS (FRESH)
1544. P.V.C COMPOUNDS (SCRAP)
1545. P.V.C ELECTRICAL INSULATING TAPE
1546. P.V.C. EXTRUSION PROFILES
1547. P.V.C FITTINGS
1548. P.V.C FLEXIBLE FUSIBLE POWDER HEAT FUSIBLE POWDER
1549. P.V.C GRANULES (FOR

INSULATION & SHEETS GRADES)
1550. P.V.C GRANULES FROM PVC SCRAPS (WITH POLLUTION CONTROL)
1551. P.V.C HOSES
1552. P.V.C LEATHER CLOTH
1553. P.V.C FLEXIBLE PIPES
1554. P.V.C PIPES AND FITTINGS
1555. POLYTHENE BAGS (PRINTED)
1556. PVC PLASTICS FILM SHEET SOFT/RIGID
1557. PVC RESIN AND COMPOUND
1558. PVC RULAR
1559. PVC STABLIZERS (SINGLE PACK SYSTEM)
1560. PVC WIRE AND CABLES
1561. PAPER BASED PHENOLIC SHEET FOR ELECTRICAL
1562. PHENOL FORMALDEHYDE RESIN
1563. PHENOLIC RESIN
1564. PLASTIC BEADS FROM PLASTIC SCRAPS
1565. PLASTIC BUTTONS
1566. PLASTIC CANS
1567. PLASTIC COLLAPSIBLE TUBE
1568. PLASTIC CORRUGATED SHEET & BOXES
1569. PLASTIC FILM AND SHEETS WITH PRINTING FLEXO & ROTO/LDPE/HDPE/PP/HM/PVC
1570. PLASTIC FILTER MASTER BATCH & OTHER MASTER BATCHES FOR VARIOUS PLASTICS
1571. PLASTIC JERRY CANS
1572. PLASTIC GRANULES OR POWDER FROM PLASTIC SCRAP
1573. PLASTIC INJECTION MOULDED T.V. CABINETS
1574. PLASTIC ITEMS MANUFAC TURE FROM POWDER MELAMINE
1575. PLASTIC GOODS
1576. PLASTIC PIPES & TARPAULINES
1577. PLASTIC SHEET FROM SCRAP
1578. PLASTIC PLANT (BLOW MOULDING & INJECTION MOULDING)
1579. PLASTIC PRODUCTS (GOLD, SILVER, NICKEL)
1580. PLASTIC TOYS
1581. POLYESTER BEADING
1582. POLYESTER FILM
1583. POLYTHENE BAGS & AUTOMATIC PRINTING
1584. POLYTHENE PRINTED BAGS
1585. POLYURETHANE FOAM AND ITS PRODUCTS
1586. PROPYLENE FILM (PRINTED) & BAG MANUFACTURING
1587. PLASTIC MATS
1588. RAINBOW COLOUR ON PVC FILM & SHEET
1589. RED MUD PVC PIPES AND FITTINGS
1590. RE-RUBBERISED OF SOLID TYPE
1591. RESIN COATED SAND REXINE
1592. REXINE CLOTH & ALLIED PRODUCTS
1593. RUBBER ADHESIVE
1594. RUBBER AUTO PARTS
1595. RUBBER BELTING
1596. RUBBER MOULDING UNIT INCLUDING LINING RUBBER SHEETING
1597. RUBBER RECLAIMING
1598. RUBBER ROLLERS FOR TEXTILE MILLS & PAPER INDUSTRIES
1599. RUBBER & PLASTIC SHEETS, MATS & FLAPS
1600. RUBBER SHEET AND ALLIED HOSPITAL RUBBER GOODS
1601. RUBBERIZED PVC GASKET
1602. RUBBERISED COIR
1603. RUBBER COMPOUND FOR AUTOMOBILES
1604. SAFETY BELTS
1605. SILVER AND GOLD PLATING ON PVC AND NYLON-6
1606. SILICONE RUBBER NIPPLES/ TEATS
1607. SPECTACLE FRAMES (PLASTIC)
1608. SPUNGE RUBBER
1609. SURGICAL EXAMINATION GLOVES
1610. SYNTHETIC PEARL COATING ON POLYSTYRENE BEADS
1611. SYNTHETIC RUBBER ADHESIVE
1612. TEFLON MANUFACTURING
1613. TEFLON TAPE
1614. TEFLON TAPES & CABLES
1615. THERMOCOLE SHEET AND MOULDED PRODUCTS EXPANDED POLYESTERENE EXTRUSION PROFILES
1616. THERMOFORMED PACKAGING (BLISTER PACKAGING & POUCH PACKAGING)
1617. THERMOPLASTIC POLYURETHANE
1618. THERMOFORMED CUPS, PLATES & GLASS WITH HIPS SHEET
1619. THERMOCOLE
1620. THERMOCOLE BASED DISPOSABLE GLASS, CUPS & PLATES
1621. TOOTH BRUSHES
1622. TOY BALLOON, DECORA TIVE & INDUSTRIAL BALLOONS
1623. TREAD RUBBER
1624. TYRE RETREADING (HOT)
1625. TYRE RETREADING BY COLD PROCESS
1626. TYRE TUBES & FLAPS
1627. TYRE & TUBES
1628. UNSATURATED POLYESTER FOR REXINE
1629. V-BELT AND FAN BELT
1630. VINYL ASBESTOS AND PVC WALL PAPER
1631. VITON (FLUORO ELASTOMER)

SOAP, COSMETICS & PERFUMS

1632. ACID SLURRY, SYNTHETIC DETERGENT POWDER
1632. ACID SLURRY
1633. AGARBATTI SYNTHETIC, PERFUMERY COMPOUND
1634. AFTER SHAVE LOTION
1635. ANTISEPTIC CREAM
1636. BETA IONONE
1637. BLEACHING POWDER
1638. BLUE DETERGENT POWDER
1639. CLEANING POWDER (VIMTYPE)
1640. COLD CREAM
1641. COLOURED FLAME & PERFUMED CANDLES (RED, BLUE, GREEN FLAME)
1642. COSMETIC INDUSTRY (MODERN)
1643. COSMETIC INDUSTRY (SHAMPOO, SPRAY PERFUME, TALCUM POWDER
1644. DETERGENT POWDER (ARIEL TYPE)
1645. DETERGENT POWDER (NIRMA TYPE) FULLY AUTOMATIC PLANT
1646. DETERGENT WASHING POWDER
1647. FISH OIL SOAP
1648. GLYCERINE TRANSPARENT SOAP
1649. HAIR REMOVING CREAM
1650. HERBAL COSMETICS
1651. HERBAL SHAMPO & CREAM
1652. HERBAL/AYURVEDIC COSMETICS
1653. INCENSE POWDER, INCENSE STICKS & INCENSE CAKE
1654. OPTICAL WHITENERS
1655. ROSE OIL EXTRACTION
1656. SHAMPOOS (COCONUT OIL BASED COLD PROCESS)
1657. SHAVING CREAM
1658. SINDUR (KUMKUM)
1659. SOAP COATED PAPER
1660. SPRAY DRIED DETERGENT POWDER
1661. STAIN REMOVER
1662. TALCUM POWDER (FACE & TOILET POWDER)
1663. TOILET AND HERBAL SOAP
1664. TOOTH PASTE & POWDER
1665. WASHING DETERGENT POWDER AND WASHING SOAP
1666. WASHING POWDER LIQUID DETERGENTS, LOTION & SHAMPOO

TEXTILE INDUSTRIES, READYMADE GARMENTS, COTTON, DYEING, BLEACHING, CLOTH, HOSIERY, POWER LOOM, HAND LOOM, WOOLLEN, SILK, SOCKS, SURGICAL COTTON & MANY OTHERS

1667. ANGORA RABBIT WOOL
1668. BUCKRAM
1669. CANVAS SHOES, JUNGLE BOOT, BOOT RUBBER KNEES & BOOT COMBAT
1670. CARPET FROM COTTON WASTE
1671. CERAMIC THREAD GUIDE
1672. COTTON BUDS/SWABS
1673. COTTON FROM WASTE YARN
1674. COTTON ROLLS
1675. COTTON SPIDERS FOR LOUD SPEAKERS
1676. DENIM CLOTH
1677. DENIM CLOTH (INTEGERATED UNIT WITH BLEACHING, DYEING AND PRINTING)
1678. DISPERSENT FOR TEXTILE
1679. DYEING & BLEACHING
1680. DYEING OF HANK YARN
1681. EMULSIFIER FOR WOOL BATCHING OIL
1682. EMBROIDERY ON FABRICS
1683. GARMENT DYEING, WASHING & STITCHING (JEANS, JACKETS, SKIRTS & SHIRTS)
1684. GOVES KNITTING
1685. GUNNY BAGS
1687. HDPE WOVEN SACKS
1688. HDPE TARPAULINS USING PLAIN LOOM WITH LAMINATION
1689. HDPE/PP WOVEN SACKS USING CIRCULAR LOOMS
1690. HOSIERY CLOTH (COTTON) PROCESSING (BLEACHING, DYEING, FINISHING OF CLOTH)
1691. HOSIERY INDUSTRY
1692. HOSIERY MERCERISING
1693. HOSIERY PRODUCTS LIKE VEST, BRIEF, T-SHIRTS & SOCKS
1694. JACQUARD FABRICS
1695. JUTE COIR, GRASS ROPE/ SUTLI MAKING
1696. JUTE FELT
1697. JUTE TWINFS
1698. KNITTED FABRICS
1699. LAMINATED JUTE BAGS
1700. LAMINATION OF HDPE WOVEN CLOTH (JUTE,COTTON, PAPER)
1701. NYLON YARN CRIMPING, DOUBLING AND BLEACHING
1702. PIGMENT BINDER FOR TEXTILE PRINTING
1704. POLYESTER RESIN FOR WIRE ENAMEL
1705. POLYESTER RESIN
1706. POLYESTER YARN FROM WASTE POLYESTER ZIP FASTENERS
1707. POWER LOOM
1708. READYMADE GARMENTS & HOSIERY
1709. READYMADE GARMENTS MERCHANDISE
1710. READYMADE GARMENTS & COVERS
1711. READYMADE GARMENTS & EMBROIDERY OF GOWNS, SHIRTS, BLOUSES, T-SHIRTS ETC. (ONLY LADIES)
1712. READYMADE SALWAR SUIT
1713. RECOVERY OF NYLON FROM NYLON WASTE
1714. ROTARY PRINTING AND DYEING ON COTTON, SYNTHETIC TEXTILE
1715. SANITARY NAPKINS
1716. SCREEN PRINTING ON COTTON CLOTH
1717. SEWING THREAD REELS & BALLS MAKING INDUSTRIES
1718. SCREEN PRINTING ON COTTON, POLYESTER & ACRYLIC
1719. SHAWLS (WOOLLEN)
1720. SHOE LACES
1721. SILK FABRICS ON HANDLOOM
1722. SOCKS KNITTING
1723. SPINNING, DOUBLING DYEING MERCERISING & BLEACHING OF COTTON YARN
1724. SPINNING & CARDING OF WOOL INTO YARNS
1725. STARCH BOOK BINDING CLOTH
1726. SURAT JARI
1727. SURGICAL COTTON & BANDAGE
1728. SYNTHETIC TEXTILE INDUSTRY(SUITING, SHIRTING, SAREES) TERRY CLOTH
1729. T-SHIRTS
1730. TOWELS, BED SHEET COVERS
1731. TERRY TOWEL
1732. TERRY FABRICS WEAVING UNIT
1733. TEXTILE AUXILLIARIES & CHEMICALS
1734. TEXTILE BLEACHING, DYEING & FINISHING & PRINTING OF COTTON FABRICS
1735. TEXTILE (HOSIERY)
1736. TEXTILE MILL
1737. TEXTILE DYEING & PRINTING
1738. TEXTILE PRINTING (JOB WORK)
1739. TEXTILE PIGMENT PRINTING BINDER
1740. TOWEL (TERRY)
1741. VELVET CLOTH BY FLOCKING PROCESS
1742. VISCOSE STAPLE FIBRE
1743. WADDING OIL (100%) FOR WADDING OF COTTON HOSIERY CLOTH IN THE DYEING PROCESS
1744. WEAVING OF DASUTI CLOTH WITH PRINTING, DYEING, EMBROIDERY & FINISHING
1745. WASHING OF JEANS & OTHER GARMENTS
1746. WORSTED WOOLLEN YARN CLOTH
1747. ZARI GLITTER SALMA SITARA OF PLASTIC FILM

INFOTECH/IT, HOSPITALITY, HOSPITAL, COLLEGE, SCHOOL, MEDICAL, ENTERTAINMENT CLUB, WARE HOUSING & REAL ESTATE, PROJECTS

1748. AMUSEMENT PARK
1749. AMUSEMENT PARK CUM WATER PARK
1750. BANQUET HALL
1751. CALL CENTER (DOMESTIC)
1752. CALL CENTER (INTERNATIONAL)
1753. CHILDREN RECREATION CENTRE
1754. COLD STORAGE
1755. COMMUNITY HALL
1756. COMPUTER EDUCATION INSTITUTE
1757. COMPUTER SOFTWARE
1758. COLLEGE
1759. CYBER CAFE
1760. DENTAL CLINIC
1761. DENTAL COLLEGE
1762. DIAGONOSTIC CENTRE
1763. E-COMMERCE/BUSINESS
1764. E-SCHOOL (Rs. 5000/-)
1765. ENGINEERING COLLEGE
1766. ENTERTAINMENT CLUB
1767. ENTERTAINMENT CLUB, HOLIDAY RESORT, 4 STAR HOTEL, AMUSEMENT PARK CUM WATER PARK, MUSHROOM & ITS PRODUCTS, FISH FARMING, LAKE FOR BOATING, DEER PARK
1768. FASHION TECHNOLOGY INSTITUTE
1769. FAST FOOD PARLOUR
1770. FIVE STAR HOTEL
1771. FOOD PARLOUR
1772. FRANCHISE TRAINING PROGRAMME FOR IIT & ENGINEERING ENTRANCE EXAMS.
1773. GOLF COURSE
1774. HEALTH CLUB, BEAUTY PARLOUR
1775. HEALTH CLUB AND FITNESS CENTER
1776. HEALTH RESORTS
1777. HOLIDAY RESORTS
1778. HOLIDAY RESORT CUM

ENTERTAINMENT CLUB WITH 4 STAR HOTEL
1779. HOSPITALS
1780. ICE CREAM PARLOUR
1781. INTERNET SERVICE PROVIDER (I.S.P.)
1782. MEDICAL COLLEGE
1783. MEDICAL COLLEGE, HOSPITAL & RESEARCH INSTITUTE
1784. MEDICAL TRANSCRIPTION CENTRE
1785. MENTAL RETARDATION HOSPITAL & CEREBRAL PALSY
1786. MOTEL/SMALL HOTEL
1787. MULTISTOREY COMMERCIAL COMPLEX
1788. MULTISTOREY RESIDENTIAL COMPLEX
1789. NURSERY SCHOOL
1790. NATURE CARE CENTRE
1791. NURSING HOME
1792. ONLINE SHOPPING MALL (RS. 5000/- REPORT)
1793. PORTAL
1794. REHABILITATION CENTRE FOR AGED & NEEDY PERSONS
1795. RESIDENTIAL CUM COMMERCIAL COMPLEX
1796. RESTAURANT
1797. RESTAURANT WITH PUB
1798. SCHOOL (PRIMARY)
1799. SCHOOL (HIGHER SECONDARY)
1800. THREE STAR HOTEL
1801. TOURIST CLUB
1802. TRAINING INSTITUTE FOR MEDICAL TRANSCRIPTION
1803. VIDEO FILM STUDIO
1804. WARE HOUSE
1805. WEBSITE DESIGN & E-MAIL REGISTERING

AGRO BASED INDUSTRIES

1806. COAL BRIQUETTES FROM AGROWASTE
1807. FURFURALFROM RICEHULL
1808. MUSHROOM CULTIVATION & PROCESSING (BUTTON)
1809. MUSHROOM GROWING & PROCESSING WITH AIR CONDITIONING
1810. MUSHROOM CULTIVATION & PROCESSING UNIT DEHYDRATION & PACKAGING OF OYSTER & PADDY STRAW MUSHROOMS
1811. ORGANIC MANURE
1812. PAPAYA CULTIVATION
1813. PAPAYA & TOMATO CULTIVATION
1814. PROCESSING & UTILISATION OF COCONUT
1815. PROCESSING OF SHEEP HAIR TO PRODUCE WOOL

BAKERY, CONFECTIONERY & FOOD PRODUCTS

1816. AGROLACTOR SOYA MILK
1817. APPLE FRUIT JUICE WITH CANNING BOTTLING
1818. AYURVEDIC SHARBAT
1819. BAKERY UNIT (PASTRIES, BREAD, BUNS, CAKE, TOFFEE ETC.)
1820. BAKING POWDER
1821. BANANA & ITS BY PRODUCTS
1822. BANANA POWDER
1823. BANANA WAFERS
1824. BANANA CULTIVATION
1825. BEER INDUSTRY
1826. DEHULLING OF JAUN FOR BEER
1827. BEER & WINE
1828. BEER, ALCOHOL, IMFL
1829. BESEN PLANT
1830. BISCUIT PLANT
1831. BREAD & BISCUIT PLANT
1832. BOTTLING PLANT COUNTRY LIQUOR FROM RECTIFIED SPIRIT
1833. BRANDY
1834. BREAD RUSKS
1835. CANNED FRUITS & VEGETABLES
1836. CANNING & PRESERVATION OF MEAT
1837. CANNING & PRESERVATION OF VEGETABLES
1838. CANNING OF MANGO PULP & MANGO SLICES
1839. CARBONATED BEVERAGS
1840. CASHEW FENI
1841. CASHEW NUT (DRIED & FRIED)
1842. CATTLE BREEDING
1843. CATTLE BREEDING & DAIRY FARM TO PRODUCE MILK
1844. CATTLE FEED FROM TAPIOCA
1845. CHEWING GINGER & AMLOKI
1846. CHEWING GUM
1847. CHILLI SAUCE
1848. CHILLI POWDER
1849. CHOCOLATE
1850. CIDER PLANT
1851. COCOA POWDER
1852. COCOA BUTTER & COCOA POWDER
1853. COCONUT SHELL POWDER
1854. COCONUT WATER
1855. COCONUT SWEET (WATERY)
1856. COCONUT MILK POWDER (DEHYDRATED)
1857. COCONUT PRODUCTS & BY PRODUCTS (INTEGRATED PLANT)
1858. COLD DRINK
1859. CURRY POWDER
1860. DAIRY FOR MILK PROCESSING (GHEE, BUTTER & PANEER)
1861. DAIRY PRODUCTS
1862. DAIRY PRODUCTS MILK PACKAGING IN POUCH (GHEE, BUTTER, ETC.
1863. DAIRY FARM TO PRDUCE MILK WITH PACKAGING (COW)
1864. DAIRY FARM TO PRDUCE MILK WITH PACKAGING (BUFALLOE)
1865. DAIRY FARM & MILK PRODUCTS
1866. DAL MOTH, CHANACHUR & BHUJIYA
1867. DEHYDRATION OF FRUITS & VEGETABLE
1868. DESICCATED COCONUT POWDER FROM COCONUTS
1869. DEHYDRATION & CANNING OF FRUITS & VEGETABLES
1870. DRY FRUIT ROASTING & PACKAGING
1871. DRYING OF RED CHILLIES, HALDI, DHANIYA, PEAS & GROUND NUT
1872. EGG POWDER
1873. FISH CANNING IN TIN & POUCHES
1874. FISH DEHYRATION (DRYING OF FISH)
1875. FISH MEAL
1876. FISH PROCESSING (BEAST FREEZING PROCESSES)
1877. FLAVOURS FOR FOOD INDUSTRIES
1878. FLOUR MILL (ROLLER)
1879. FROZEN MEAT
1880. FOOD DEHYDRATION (FRUITS & VEGETABLES)
1881. FRIED & ROASTED GROUND NUT, GRAMS, PEAR ETC.
1882. FRUIT JUICE MAKING & PACKAGING IN PLASTIC CONTAINER
1883. FRUIT JUICE IN TETRA PACK (DRINKS)
1884. FRUIT JUICE, SQUASHES, SAUCE & KETCHUP, JAM, JELLY, VINEGAR ETC.
1885. GARLIC FLAKES
1886. GARLIC POWDER
1887. GHEE & BUTTER
1888. GINGER (PULVERISED)
1889. GINGER GLAZING & PRESERVATION
1890. GINGER STORAGE
1891. GINGER PROCESSING
1892. GINGER OIL & GINGER DUST
1893. GINGER POWDER (DRY) & OLEORESIN
1894. GRAPE DEHYDRATION
1895. GRAPE CULTIVATION
1896. GRAPE JUICE
1897. HONEY PROCESSING & PACKAGING
1898. ICE CREAM & ICE CANDY
1899. ICE CUBE
1900. INVERT SUGAR
1901. IDLI MIX, DOSA MIX SAMBHAR MIX, VADA MIX, GULAB JAMUN MIX
1902. INSTANT COFFEE & INSTANT TEA

1903. INSTANT NOODLES
1904. INSTANT SOUPS
1905. IODIZED SALT FROM CRUDE SALT
1906. JAM CHUTNEY PICKLES & SQUASHES
1907. KATHA MANUFACTURING
1908. LECITHIN (SOYA BASED)
1909. LEMON & ITS PRODUCTS
1910. MACARONI MANUFACTURING
1911. MACARONI, SPAGHETTI, VERMICELLI & NOODLES
1912. MAIZE & ITS BY PRODUCTS MALTING PLANT
1913. MANGO PAPPAD (AAM PAPPAD)
1914. MANGO POWDER RIPE
1915. MANGO POWDER
1916. MANGO PROCESSING & CANNING (MANGO PULP)
1917. MEAT PROCESSING (CHICKEN MUTTON)
1918. MEAT PROCESSING (BUFFALO)
1919. MENTHOL BOLD FROM MENTHOL FLAKES
1020. MILK POWDER
1921. MILK POWDER & GHEE
1922. MILK POWDER, GHEE & SPICES
1923. MILK PRESERVATION & MARKETING TO WHOLE SELLERS
1924. MILK PRESERVATION & MARKETING TO WHOLE SELLERS (IN POUCHES)
1925. MILK TOFFEE MANUFACTURES
1926. MINERAL WATER
1927. MINERAL WATER IN POUCHES
1928. MINI FLOUR MILL ATTA MAIDA, SUJI & WHEAT BRAN
1929. MITHAI/HALWAI (SWEET & NAMKEEN)
1930. MUTTON PROCESSING
1931. PAN MASALA (MEETHA, SADA, ZARDA) MAKING & PACKING
1932. PAN MASALA AND POUCH MAKING
1933. PANEER (CHEESE)
1934. PAPAD & BARIYAN
1935. PAPAD PLANT
1936. PEPSICOLA IN POLYTUBES
1937. PETHA PACKAGING
1938. PICKLES
1939. PIGGERY/MEAT/CHICKEN PROCESSING
1940. PINE APPLE JUICE CANNING
1941. POTATO & ONION POWDER
1942. POTATO & ONION FLAKES
1943. POTATO CHIPS
1944. POTATO GRANULES
1945. POUCH FILLING FOR SAUNF SUPARI ILAICHI ETC.
1946. PRESERVATION OF RAWS MANGO JUICE
1947. PROCESSED CHEESE & MARINE PDTS.
1948. PULP FROM TARMARIND
1949. READYMATE PROCESSED FOOD
1950. RICE & CORN FLAKES
1951. RICE BASMATI (TRADING)
1952. RICE POLISHING & PACKAGING IN POUCH
1953. ROASTED/SALTED/MASALA/ CASHEW NUTS ALMONDS & PEA NUT
1954. ROLLER FLOUR MILL

INFOTECH/IT PROJECTS

1955. MEDICAL TRANSCRIPTION
1956. E-COMMERCE
1957. CYBERCAFE
1958. INTERNET SERVICE PROVIDER (ISP)
1959. COMPUTER EDUCATION CENTRE
1960. PORTAL (WEBSITE DESIGN)
1961. CALL CENTRE

TERMS AND CONDITIONS

FOR INDIA

1. Please ask for Quotations of Each *Market Survey Cum Detailed Techno Economic Feasibility Report.* Payable in advance through Draft/M.O./Cash in favour of 'ENGINEERS INDIA RESEARCH INSTITUTE', Delhi. Delivery by Regd.post within 2 Days. (Postage Free)

2. Ask for the Hi-Tech Projects Monthly magazine. Each Magazine covers many Project Profiles for selection of Industry.

FOR OVERSEAS

1. Please ask for Quotations in US $ of Each *Market Survey Cum Detailed Techno Economic Feasibility Report.* Payable in advance through Draft/M.O./Cash in favour of 'ENGINEERS INDIA RESEARCH INSTITUTE', Delhi. (Payable in India) Delivery by Regd.post within 4 Days.

2. Ask for the Hi-Tech Projects Monthly magazine. Each Magazine covers many Project Profiles for selection of Industry.

ENGINEERS INDIA RESEARCH INSTITUTE

4449, Nai Sarak (D), Main Road, Delhi - 110 006 (India)
Ph:. 91-11-23918117, 23916431, 23840361, 31082370
Fax: 91-11-23916431 E-Mail : eirisidi@bol.net.in
Website:www.startindustry.com
Also at : 4/27, Roop Nagar, Near Roop Nagar No. 1 School, Delhi - 110 007 (India) Ph:. 23840361